**Louis de Broglie**

# Research on the Theory of Quanta

With a Foreword by Hirokazu Nishimura

Translated by André Michaud and Fritz Lewertoff

Edited by Vesselin Petkov

Louis de Broglie
15 August 1892 – 19 March 1987

Book cover photo: https://commons.wikimedia.org/wiki/File:Broglie_Big.jpg



ISBN 978-1-927763-98-8 (softcover)
ISBN 978-1-927763-99-5 (ebook)

Minkowski Institute Press
Montreal, Quebec, Canada
minkowskiinstitute.org/mip/

For information on all Minkowski Institute Press publications visit our website at minkowskiinstitute.org/mip/books/

# Preface

*Louis de Broglie's Nobel Lecture contains a nice reminder, which, although regarded as self-evident in physics, has recently become unexpectedly relevant, that: "experiment... is the final judge of theories" p. 132.*

This volume contains the long overdue first publication in English of Louis de Broglie's 1924 dissertaion *Recherches sur la théorie des Quanta*[1] and his 1929 Nobel lecture "The Wave Nature of the Electron."[2]

De Broglie's dissertation was typeset into LaTeX, but the original figures were intentionally kept. A number of Editor's Notes were added in the text and noticed typos were corrected.

I would like to thank Hirokazu Nishimura for writing the Foreword, which makes this publication even more valuable.

I would also like to thank André Michaud and Fritz Lewertoff for the translation (as always on a voluntary basis) of de Broglie's dissertation as their continuous contributions to the missions of the *Minkowski Institute* and the *Minkowski Institute Press.*

I would like to express my gratitude to my wife Svetla Petkova for the hard work of typesetting de Broglie's translated text into LaTeX.

Montreal
18 January 2021

*Vesselin Petkov*
*Minkowski Institute*

---

[1] Louis de Broglie, *Recherches sur la théorie des quanta,* Thesis, Paris, 1924, *Ann. de Physique* (10) 3, 22 (1925).

[2] Louis de Broglie – Nobel Lecture. NobelPrize.org. Nobel Media AB 2021. Sun. 17 Jan 2021. https://www.nobelprize.org/prizes/physics/1929/broglie/lecture/

# Foreword by Hirokazu Nishimura

De Broglie (1892-1987) was a French physicist making groundbreaking contributions to burgeoning quantum theory and establishing *wave mechanics* to be completed by Erwin Schrödinger (1887-1961). His 1924 thesis entitled Recherches sur la théorie des quanta (Research on the Theory of Quanta), translated from French into English in this book, established the wave-particle duality theory of matter, expounding his firm conviction, known as the *de Broglie hypothesis,* that any moving particle had an associated wave. The thesis was intended for his doctoral degree from the examining board of the Sorbonne consisting of Jean Perrin, Paul Langevin, Elie Cartan and Charles Maugin. The examining board, perplexed by apparently radical ideas of de Broglie, asked Albert Einstein (1879-1955) whether the thesis deserved a doctoral degree. Einstein responded quickly by saying that the thesis deserved a Nobel Prize rather than a doctoral degree. Einstein recommended the thesis to Schrödinger, which resulted in celebrated Schrödinger equation. De Broglie, after his theory was verified experimentally by G. P. Thomason's thin metal diffraction experiment of passing a beam of electrons through thin film of celluloid on the one hand and by Davisson and Germer's experiment of guiding their beam through a crystalline grid on the other in 1927, was awarded a Nobel Prize in Physics in 1929. One of the central results of de Broglie's thesis is that, with mass $m$, velocity $v$, and momentum $p$, we have

$$\lambda = \frac{h}{p} = \frac{h}{mv}\sqrt{1-\frac{v^2}{c^2}}$$

where $c$ is the speed of light in vacuum, $h$ is Planck's constant and $\lambda$ is the wavelength. The distinction between phase velocity and group velocity is significant, because we should use the former in discussing wavelength and frequencies. Here I should refer to Edward MacKinnon's critical comment [De Broglie's thesis: A critical retrospective, American Journal of Physics, **44** (1976)], which claims that de Broglie's original development rested on confusion between the velocity of a particle and the relativistic phase wave de Broglie associated with it on the one hand and on confusion between the group velocity of a wave packet and the velocity of individual waves in the packet on the other, though de Broglie arrived finally at the well-known right formula.

The wave theory of matter by de Broglie was preceded by that of light. It is said that Euclid (possibly 330 B.C.-275 B.C.) found out the first and second laws of geometric optics. The third law of geometric optics called Snell's law was found out accurately by the Persian scientist Ibn Sahl in 984, being rediscovered by Thomas Harriot in 1602 and being rediscovered again by Dutch astronomer Willebroad Snellius (1580-1626) in 1621. During the 17th and 18th centuries Issac Newton's (1642-1727) corpuscular theory of light, according to which light is emitted from a luminous body in the form of tiny particles, was dominant, though many scientists such as Robert Hooke, Christian Huygens and Leonhard Euler proposed a wave theory of light based on experimental observations, only to be eclipsed by the highest stature of Issac Newton. Thomas Young's famous double-slit experiment at the beginning of the 19th century played a major role in the general acceptance of the wave theory of light. Maxwell's famous equations for classical electromagnetism, discovered by James Clerk Maxwell (1831-1879) in the middle of the 19th century, gave a decisive support to the wave theory of light, meaning that light is no other than a electromagnetic radiation (fluctuating electric and magnetic fields propagating at a constant speed in vacuum).

The old quantum theory began on 14 December of 1900 when Max Planck (1858-1947) proposed the central assumption on black-body radiation, known as the *Planck postulate*, that electromagnetic energy could be emitted only in quantized form, the energy being only a multiple of an elementary unit

$$E = h\nu$$

where $h$ is Planck's constant and $\upsilon$ is the frequency of radiation. In 1905 Albert Einstein explained why the maximum kinetic energy of the outgoing electrons depends on the light frequency rather than on its intensity by depicting light as composed of discrete quanta, now called *photons*, rather than continuous waves and theorizing the above formula, which brought Einstein a Nobel prize in physics in 1921 after Einstein's equation incorporating photons was verified by the American experimenter Robert A. Millikan. Einstein's theory had a great influence on de Broglie and his thesis. Einstein's seemingly far-fetched idea of photons was crazy enough to take several years to be accepted generally, because the wave theory of light was firmly established at that time. In 1905 Einstein also established the equivalence between mass and energy

$$E = mc^2$$

which also had a great influence on de Broglie and his thesis.

*Matrix mechanics* was the first conceptually autonomous system of quantum mechanics formulated by Werner Heisenberg (1901-1976), Max Born (1882-1970) and Pascual Jordan (1902-1980) in 1925. Matrix mechanics is not easy to handle. Physicists succeeded in calculating the spectrum of a hydrogen atom by matrix mechanics, but they could not proceed to the next stage of calculating the spectrum of a helium atom. Then came wave mechanics, which gave physicists great joy, because they could return from their unfamiliar discrete world to their continuous homeland. The birth of wave mechanics irritated Heisenberg and some others greatly. However, Wolfgang Ernst Pauli (1900-1958) soon showed the equivalence between matrix mechanics and wave mechanics, Schrödinger arriving at the same conclusion a month later in a bit incomplete way.

It is a familiar episode that Richard Feynman (1918-1988) said

If you think you understand quantum mechanics,
you don't understand quantum mechanics.

This statement is like a word of a Zen Buddhist. This means that we have a mathematical formalism well adequate for quantum theory predicting exactly what quantum phenomena will take place, while you can not or should not pursue the meaning of the formalism nor consider what the formalism represents physically. Many great physicists like Planck, Einstein or de Broglie, who were raised thoroughly in classical physics and contributed much to the old quantum theory, could not content themselves with such a practical attitude as the Copenhagen interpretation proposed by Niels Bohr (1885-1962) and Werner Heisenberg from 1925 through 1927, which could not relieve them at all. Perplexed by his own discovery in 1900, Plank felt as if he were Epimetheus, the Titan in Greek mythology who opened Pandora's box. Indeed, Planck seriously attempted in vain to erase quantum theory from the world. Einstein attacked quantum theory by famous paradoxes. De Broglie tried to build a comprehensible quantum theory. In the 1920s de Broglie was vigorously engaged in the *pilot wave model*, and his early considerations on it were already in his thesis. In addition to a wave function on the space of all possible configurations, the pilot wave model postulates an actual configuration existing even when unobserved. The evolution of the configuration over time is determined by the guiding equation which is a nonlocal part of the wave function. The evolution of the wave function over time is provided by the Schrödinger equation. De Broglie was persuaded to give up his pilot wave model in favor of the then mainstream Copenhagen interpretation. David Bohm (1917-1992) rediscovered de Broglie's pilot wave model in 1952, so that the pilot wave model is usually called the *de Broglie-Bohm theory*. The theory is the first known example of a *hidden variable theory*, interpreting quantum mechanics as a deterministic theory and avoiding troublesome notions such as wave-particle duality, instantaneous wave function collapse and the paradox of Schrödinger's cat. The price that the theory should pay for such benefits is its inherent nonlocality. By Bell's theorem no *local* hidden variable theory is possible.

The second half of the thesis was devoted to making use of the equivalence between the mechanical principle of least action and Fermat's optical principle, which resulted in

- Fermat's principle applied to phase waves is identical to Maupertuis's principle applied to the moving body.
- The possible dynamic trajectories of the moving body is identical to the possible rays of the wave.

The thesis was translated into German and published in 1927 (Untersuchungen zur Quantentheorie, translated from French into German by W. Becker. Akademische Verlagsgesellschaf). It is interesting to note that it took only three years for its German translation to be published, while it took almost a century for its English translation to be published. This reminds us of supremacy of the Weimar culture (recall, e.g., Vienna circle, including Moritz Schlick, Hans Hahn, Kurt Gödel, Rudorf Carnap, Richard von Mieses, and so on as regular members, occasionally visited by Alfred Tarski, Hans Reichenbach, Oskar Morgenstern, Willard van Orman Quine, Frank F. Ramsey, and usually in close contact with Karl Popper and Ludwig Wittenstein on the one hand, and the Berlin circle created by Hans Reichenbach, Kurt Grelling and Walter Dubislav and composed of Carl Gustav Hempel, David Hilbert Richard von Mises on the other, the two circles publishing the journal Erkenntnis or Knowledge in English).

Hirokazu Nishimura (Tsukuba, Japan)

## Preface by Louis de Broglie

Louis de Broglie added this preface in the 1963 republication of the 1924 text by Masson et Cie, Editeurs, Paris.

As I leave the Chaire de Théories Physiques de la Faculté des Sciences de Paris and this Institut Henri Poincaré where for 34 years I have been teaching the theories of contemporary physics, trying to orient the work of young researchers, a group of former students or disciples, several of whom now occupy important positions in the university hierarchy or at the Centre National de la Recherche Scientifique, took the initiative to reprint the doctoral thesis that I defended in front of the Faculté des Sciences de Paris on November 25, 1924, a work that can be considered as having been at the origin of all of the development of the theories that form what we now refer to as "Wave Mechanics" or "Quantum Mechanics."

I am extremely moved by the outpouring of sympathy conveyed by the reprinting of my Thesis and I want first and foremost to address to all those who have taken this initiative, to all those who, in the past or more recently, directly or indirectly, have been my collaborators or my students, my feelings of deep gratitude and friendly devotion.

To see this already ancient text reprinted today draws my attention back to a distant episode in my life. Such a recollection is at the same time sweet and a little sad: it evokes the enthusiasm and the wonderful horizons of youth, but it also reminds me of the swift passage of time so poetically expressed by Malherbe when he wrote:

> The pleasure of days is in their morning
> Night is already close for one who crosses midday.

I thus recall those brief years when my thoughts, nourished by innumerable readings extending to the most diverse fields, kept coming back to the serious issue of the double granular and ondulatory aspect of Light that Einstein had raised nearly twenty years earlier, in his brilliant Theory of light quanta. After having pondered at length in solitude and meditation, the idea suddenly occurred to me in 1923, that the discovery made by Einstein in 1905 should be generalized to all material particles, and in particular to electrons. I then discovered the relations that generalize those of the theory of light quanta, that allow establishing, between any material particle and the wave that I associated with it, a relation analogous to the one Einstein established between the electromagnetic wave and what we now call the photon. Presented in notes to the proceedings of the Académie des Sciences in the early fall of 1923, these ideas, somewhat developed and clarified, were the subject of my Thesis the following year.

As I boldly ventured into entirely unexplored territory, I was convinced, as texts published during this period clearly show, that it was essential to achieve a true synthesis of the notions of waves and corpuscles while preserving the precise images of physical realities that always were attached to these two notions. During the three years after the submission of my Thesis, I tried to construct, in a rather imperfect way I must admit, a theory in conformity with the goal that I had set out to achieve. I do not have to recall here, as I have done several times elsewhere in recent years, how at the time of the Solvey Conference of October 1927, my theory met with strong opposition, and how, discouraged by the objections that were made, I rallied, with perhaps still some reluctance, to the very different interpretation that was opposed to mine by eminent scholars and which has remained since then the "orthodox" interpretation of Quantum Mechanics. This interpretation, due especially to the so-called "Copenhagen School" was never accepted, it must be remembered, by great promoters of the Physics of Quanta, such as Max Planck, Albert Einstein or Erwin Schrödinger who, until their death, never stopped opposing it.

About ten years ago, after some new in-depth reflections on this capital problem of contemporary Physics, I reverted to the idea of my attempts of 35 years ago, which, as insufficient as they may have been, were nevertheless oriented in the right direction. Careful examination of the objections made to this orthodox theory by the illustrious physicists that I just mentioned, and by a few others, led me to conclude that this interpretation leads to some paradoxical and truly unacceptable conclusions, that are hidden under an elegant and precise mathematical formalism whose main lines I know well for having studied and taught it at length. In my opinion, these paradoxical conclusions all stem from the fact that the clear image of the corpuscle as a concentration of energy closely localized in space over time has been abandoned and that, abandoning the idea that the wave is a physical process actually propagating in space, the wave function has come to be considered as a simple mathematical artifice for calculating probabilities.

So I resumed my former attempt at considering the corpuscle as a very small region of high field concentration incorporated in an extended wave, but I introduced new elements that I think have improved it a bit. First, I recognized more and more clearly the need, according to my conception, to introduce non-linearity in the wave equations of Wave Mechanics because only strong and very localized non-linearity seems to me to be able to account for the existence of a strong and permanent concentration of energy of the desired type within a wave obeying everywhere else a substantially linear equation. Besides, this introduction of non-linearity also seemed more and more to me as an appropriate means to achieve one day a physical interpretation of what we refer to as quantum transitions. Quantum transitions include not only those that, in atomic and molecular systems, accompany the emission and absorption of radiation, but also all exchanges of energy and momentum between observable particles or systems. The orthodox interpretation asserts that these transitions completely and definitively escape all our representation modes, which seems to me to constitute a total abandonment of any hope of rational explanation. It seems to me more satisfactory to admit that quantum transitions are very fast transient processes probably of a non-linear character, the description of which escapes current linear theories, but would probably become possible within the framework of a broader non-linear theory. With the help of a few young collaborators[3] whom I must particularly thank for their sympathy and help, I have been thinking a lot lately about the new perspectives that are opened up by this conception of quantum transitions and I had the impression of foreseeing vast new horizons.

Another idea that I thought needed to be added to the grounding conditions of my double solution theory concerns the introduction of a random element in the motion of the corpuscle within its wave. One of the essential characteristics of the development of Wave Mechanics was, in fact, to introduce the concept of probability even when dealing with a single particle apparently isolated. Within the framework of the ideas I am adopting, that reverts to the traditional interpretation of the concept of probability, this is hardly conceivable unless the apparently isolated particle is in reality in constant interaction with a very complex hidden environment. We thus reconnect with the hypothesis put forward in 1954 by Bohm and Vigier that what we call the vacuum would be the seat of a sub-quantum medium, a sort of immense pool of hidden energy, of which the world of microphysical particles would in some way only constitute the observable part. It would undoubtedly be premature to attempt drawing a precise picture of such a sub-quantum environment, but the mere fact of admitting its existence leads one to compare any observable particle to a granule in energetic contact with a hidden thermostat analogous to the granules in Jean Perrin's memorable experiments. It therefore seems natural to introduce, even in the study of a single, apparently isolated parti-

[3]MM. Francis Fer, Georges Lochak and João Luis Andrade e Silva.

cle, notions of statistical thermodynamics that allow the use of methods that have long been classical in the theory of fluctuations. This is what I tried to do recently, and although this attempt is still a rather vague outline, it too seems to me to offer a glimpse of very broad horizons.

It is essential to note that my efforts to develop a new interpretation of Wave Mechanics are in no way meant to challenge the accuracy of the predictions of Quantum Mechanics or Quantum Field Theory. In agreement with a statement of Einstein that I have often quoted, I think that the usual theory allows accurate statistical predictions, but that it does not provide a complete description of physical reality, which must be expressed by a precise synthetic image of the wave and the corpuscle and their mutual relations. I will be opposed with statements to the effect that my reinterpretation is useless because, I will be told, it has not led to any new predictions of observable phenomena. I admit that this is currently correct (except perhaps with respect to the general particle theory of M. Vigier and his collaborators in the framework of space-time). But I believe that I can counter the previous objection with the following two remarks. First, if an interpretation of the type I envision can be developed on solid foundations, it will allow understanding the true nature of the waves and corpuscles dualism, which is in no way possible with the very abstract formalisms currently in vogue, nor with the rather obscure notion of complementarity. Now, I think that the goal of Science at its highest level has always been to understand. Moreover, the history of Science shows that whenever we were able to fully understand the true nature of a class of physical phenomena, this better understanding always quickly led to new predictions and applications.

The results of my efforts are still, I confess, rather fragmentary. Nevertheless, as my Thesis is being republished, my thoughts are brought back to the time when I was the first to glimpse in all its generality the problem of the relation between waves and corpuscles and I think I can say that this problem is far from being completely solved by the orthodox interpretation of Quantum Mechanics. I must therefore conclude by hoping that more young researchers will turn to the study of this exciting problem, drawing directly from the data of experience, freeing themselves from preconceived ideas and not placing too exclusive a value on mathematical formalisms, even elegant and rigorous ones, which sometimes can hide deep physical realities.

10 December 1962,<br>Louis de Broglie.

# CONTENTS

**Preface by the Editor** **i**

**Foreword by Hirokazu Nishimura** **iii**

**Preface by Louis de Broglie** **ix**

**Summary** **1**

**Historical Introduction** **3**
I. From the 16th to the 20th Century. . . . . . . . . . . . . . 3
II. The 20th century: Relativity and Quanta . . . . . . . . . 5

**1 The Phase Wave** **11**
I. The Relation between Quantum and Relativity . . . . 11
II. Phase Velocity and Group Velocity . . . . . . . . . . . . 17
III. The Phase Wave in Space-Time . . . . . . . . . . . . . . 19

**2 Principle of Maupertuis and Principle of Fermat** **27**
I. Purpose of this Chapter . . . . . . . . . . . . . . . . . . . . 27
II. The Two Principles of Least Action in Classical Dynamics . . . . . . . . . . . . . . . . . . . . . . . . . . . . 29
II. The Two Principles of Least Action in the Dynamics of the Electron . . . . . . . . . . . . . . . . . . . . . . . . . 31
IV. Wave Propagation; Fermat Principle . . . . . . . . . . . 35
V. Extension of the Quantum Relation . . . . . . . . . . . . 37
VI. Specific Cases; Discussions . . . . . . . . . . . . . . . . . 39

**3 The Quantum Conditions of Trajectory Stability** **49**
1. The Bohr-Sommerfeld Stability Conditions . . . . . . 49
II. Interpretation of Einstein's Condition . . . . . . . . . 51
III. Sommerfeld Conditions for Quasi-Periodic Motions 52

**4 Quantification of the Simultaneous Motions of two Electrical Centers** **57**
I. Difficulties Raised by this Issue . . . . . . . . . . . . . . 57
II. The Drag of the Nucleus in the Hydrogen Atom . . . . 58
III. The two Phase Waves of the Nucleus and the Electron 61

**5 The Quanta of Light** **65**
I. The Atom of Light . . . . . . . . . . . . . . . . . . . . . . . . . 65
II. The Motion of the Atom of Light . . . . . . . . . . . . . . 69
III. Some Agreements between the Opposing Radiation Theories . . . . . . . . . . . . . . . . . . . . . . . . . 70
IV. Wave Optics and Light Quanta . . . . . . . . . . . . . 75
V. Interferences and Coherence . . . . . . . . . . . . . . . . . 76
VI. The Bohr Frequency Law. Conclusions . . . . . . . . 78

**6 The Scattering of $X$– and $\gamma$-rays** **81**
I. The Theory of J. J. Thomson . . . . . . . . . . . . . . . . . 81
II. The Theory of Debye . . . . . . . . . . . . . . . . . . . . . . . 84
III. The Recent Theory of P. Debye and A. H. Compton . 86
IV. Scattering by Moving Electrons . . . . . . . . . . . . . . 90

**7 Statistical Mechanics and Quanta** **95**
II. Reminder of Some Results of Statistical Thermodynamics . . . . . . . . . . . . . . . . . . . . . . . . . 95
II. New Conception of the Statistical Equilibrium of a Gas 100
III. The Gas of Light Atoms . . . . . . . . . . . . . . . . . . . 105
IV. Energy Fluctuations in the Black Body Radiation . . . 110

**Appendix to Chapter 5**
**On Light Quanta** **113**
Summary and Conclusions . . . . . . . . . . . . . . . . . . . 116

THE WAVE NATURE OF THE ELECTRON
**Louis de Broglie's Nobel Lecture, December 12, 1929** 119

# Summary

The history of optical theories shows that scientific thought has long hesitated between a dynamic and an undulatory conception of light; these two representations are therefore probably less in opposition than was previously assumed, and the development of quantum theory seems to confirm this conclusion.

Guided by the idea of a general relationship between the notions of frequency and energy, we admit in the present work the existence of a periodic phenomenon of an yet to be specified nature that would be related to any isolated amount of energy and that would depend on its own mass by means of the Plank-Einstein equation. The relativity theory then leads to associate the propagation of a certain wave to the uniform motion of any material point, whose phase moves in space faster than light (ch. I).

To generalize this result to the case of non-uniform motion, we are led to admit a relation of proportionality between the Universe Impulse vector of a material point and a vectorial characteristic of propagation of the associated wave whose time component is the frequency. The Fermat principle applied to the wave then becomes identical to the principle of least action applied to the moving object. The rays of the wave are identical to the possible trajectories of the mobile (ch. II).

The previous statement applied to the periodic motion of an electron in the Bohr atom allows us to find the quantum stability conditions as expressions of the resonance of the wave along the length of the trajectory (ch III). This result can be extended to the case of circular motions of the nucleus and the electron around their common center of gravity in the hydrogen atom (ch. IV).

The application of these general ideas to the light quantum conceived by Einstein leads to many very interesting correlations. Despite the difficulties that remain, it allows us to hope for the constitution of an optic that could be both atomistic and undulatory, establishing a sort of statistical correspondence between the wave linked to the light energy quantum and Maxwell's electromagnetic wave (ch. V).

In particular, the study of the scattering of $X$-rays and $\gamma$-rays by amorphous bodies serves to show how much such a correspondence is desirable today (ch. VI).

Finally, the introduction of the notion of a phase wave in statistical mechanics leads to justify the introduction of quanta in the dynamic theory of gases and to rediscover the laws of black body radiation as reflecting the distribution of energy between atoms in a light quanta gas.

# Historical Introduction

## I. From the 16th to the 20th Century

Modern science was born at the end of the 16th century as a result of the intellectual revival due to the Renaissance. While positional astronomy became more precise by the day, the sciences of equilibrium and motion, statics and dynamics slowly came into being. It is well known that it was Newton who first made Dynamics a homogeneous body of doctrine and by means of his memorable law of universal gravitation opened up an enormous field of applications and validations for the new science. During the 18th and 19th centuries a great number of geometricians, astronomers and physicists developed Newton's principles, and Mechanics reached such a degree of beauty and rational harmony that its character as a physical science was almost forgotten. In particular, it was possible to derive all this science from a single principle, the principle of least action, first stated by Maupertuis, and then in another way by Hamilton, whose mathematical form is remarkably elegant and condensed.

Through its involvement in Acoustics, Hydrodynamics, Optics, Capillarity, Mechanics seemed for a moment to reign over all fields. It had a little more difficulty integrating a new branch of science born in the 19th century: Thermodynamics. If one of the two great principles of this science, that of energy conservation, is easily interpreted by the concepts of Mechanics, the same cannot be said of the second, that of the increase in entropy. The work of Clausius and Boltzmann on the analogy of thermodynamic quantities with certain quantities involved in periodic motion, work that is nowadays back on the agenda, did not succeed in completely restoring the agreement of these two points of view. However, the admirable kinetic theory of gases by Maxwell and Boltzmann and the more general doctrine of Statistical Mechanics by Boltzmann and Gibbs showed that Dynamics, if complemented by considerations of probability, allows the interpretation of the fundamental notions of thermodynamics.

As early as the 17th century, the science of light, optics, had attracted the attention of researchers. The most common phenomena (rectilinear propagation, reflection, refraction), those which today form our geometric optics, were naturally the first known. Several scientists, including Descartes and Huyghens, worked to unravel their laws and Fermât summarized them in a synthetic principle that bears his name and which, stated in our current mathematical language, recalls by its form the principle of least action. Huyghens had leaned towards a wave theory of light, but Newton, sensing in the great laws of geometric optics a deep analogy with the dynamics of the material point of which he was the creator, developed a corpuscular theory of light termed “emission theory” and even managed to account for phenomena now classified in wave optics with the help of somewhat artificial hypotheses. (Newton’s rings).

The beginning of the 19th century saw a reaction against Newton’s ideas in favour of those of Huyghens. Interference experiments, the first of which were due to Young, were difficult if not impossible to interpret from a corpuscular viewpoint. Fresnel then developed his admirable elastic theory of light wave propagation, and from then on the credit for Newton’s conception steadily diminished.

One of Fresnel’s great successes was to explain the rectilinear propagation of light, the interpretation of which was so intuitive in emission theory. When two theories, based on ideas that seem completely different to us, can account with the same elegance for the same experimental truth, one can always wonder whether the opposition between the two points of view is real or solely due to the insufficiency of our efforts at synthesising them. This issue did not come up in Fresnel's time, and the concept of light consisting of corpuscules was considered naïve and abandoned.

The 19th century saw the birth of a brand new branch of physics that revolutionized our world view and our industry: the science of Electricity. There is no need to recall here how it was developed thanks to the work of Volta, Ampère, Laplace, Faraday, etc. The only important point is to mention that Maxwell was able to summarize the results of his predecessors in formulas of superb mathematical conciseness and to show how optics as a whole could be considered a branch of electromagnetism. The work of Hertz and even more so that of Mr H. A. Lorentz perfected Maxwell's theory; Lorentz also introduced the notion of the discontinuity of electricity already developed by Mr J. J. Thomson and so brilliantly confirmed by experiment. It is true that the development of the electromagnetic theory deprived the Fresnel elastic ether of its reality and thus seemed to separate optics from the field of Mechanics, but many physicists, like Maxwell himself, still hoped at the end of the last century to find a mechanical explanation of the electromagnetic ether and, as a result, not only to subject optics once again to dynamic explanations, but also to subject all electrical and magnetic phenomena to it at the same time. The past century thus ended enlightened by the hope of a forthcoming and complete synthesis of all physics.

## II. The 20th century: Relativity and Quanta

However, there were still some shadows on the horizon. Lord Kelvin, in 1900, announced that two dark clouds seemed to be threatening on the horizon of Physics. One of these clouds represented the difficulties raised by Michelson and Morley's famous experiment, which seemed incompatible with the ideas accepted at the time. The second cloud represented the failure of the methods of Statistical Mechanics in the field of the black body radiation; the theorem of equipartition of energy, a rigorous consequence of Statistical Mechanics, leads indeed to a well defined distribution of energy between the various frequencies in the radiation of thermodynamic equilibrium; however, this law, the Rayleigh-Jeans law, is in gross contradiction with experiment and it is even almost absurd because it foresees an infinite value for the total energy density, which obviously makes no physical sense.

In the early years of the twentieth century, Lord Kelvin's two clouds condensed, so to speak, one into the theory of Relativity and the other into the Quantum theory.

How the difficulties raised by Michelson's experiment were first studied by Lorentz and Fitz-Gerald, how they were then resolved by Mr A. Einstein thanks to an intellectual effort perhaps without example, this is what we will not develop here, the question having been exposed many times in recent years by voices more authoritative than ours. We will therefore assume in this paper that the essential conclusions of the theory of Relativity are known, at least in its special form, and we will make use of them whenever appropriate.

We will, on the contrary, quickly focus on the development of quantum theory. The notion of quanta was introduced into science in 1900, by Mr. Max Planck. This scientist was then theoretically studying the question of the black body radiation and, since the thermodynamic equilibrium should not depend on the nature of the emitters, he conceived a very simple emitter called "Planck's resonator" made up of an electron subjected to a quasi-elastic bond and thus possessing a vibration frequency independent of its energy. If we apply the classical laws of Electromagnetism and Statistical Mechanics to the exchange of energy between such resonators and radiation, we find Rayleigh's law, whose undeniable inaccuracy we mentioned earlier. In order to avoid this conclusion and to find results more in line with experimental facts, Mr. Planck admitted a strange postulate: "The exchanges of energy between resonators (or matter) and radiation take place only in finite quantities equal to $h$ times the frequency, $h$ being a new universal constant in physics". To each frequency, therefore, corresponds a sort of energy atom, a *quantum* of energy. The observation data provided Mr. Planck with the necessary basis for the calculation of the constant $h$ and the value found then ($h = 6.545 \times 10^{-27}$) was hardly modified by the innumerable subsequent determinations made by the most diverse methods. This is one of the most beautiful examples of the power of theoretical physics.

The concept of quanta quickly spread and soon permeated every part of Physics. Whereas their introduction removed some of the difficulties relative to the specific heats of gases, it allowed Mr. Einstein first, then Messrs. Nernst and Lindemann, and finally in a more perfect form Messrs. Debye, Born and von Karman, to develop a satisfactory theory of the specific heats of solids and to explain why the Dulong and Petit law sanctioned by classical statistics has important exceptions and is, like Rayleigh's law, only a valid limiting form in a certain domain.

Quanta also broke into a science where they were hardly expected: gas theory. Boltzmann's method leads to leave undetermined the value of the additive constant in the expression of entropy. Mr. Planck, in order to account for Nernst's theorem and to obtain the exact prediction of the chemical constants, admitted that it was necessary to make quanta intervene and he did it in a rather paradoxical manner by attributing to the element of extension in phase of a molecule a finite quantity equal to $h^3$.

The study of the photoelectric effect raised a new conundrum. A photoelectric effect is the expulsion of moving electrons from matter under the influence of radiation. Paradoxically, experiments show that the energy of the electrons expelled depends on the frequency of the excitation radiation and not on its intensity. Mr. Einstein, in 1905, accounted for this strange phenomenon by admitting that radiation can only be absorbed by quanta of value $h\nu$; therefore, if the electron absorbs energy $h\nu$ and if it must work $\omega$ to get out of matter, its final kinetic energy will be $h\nu - \omega$. This law has been well verified. With his deep intuition, Mr. Einstein felt that it was necessary to return in some way to the corpuscular conception of light and hypothesized that all radiation of frequency $\nu$ is divided into atoms of energy of value $h\nu$. This light quantum hypothesis (Lichtquanten) in opposition to all the facts of Wave Optics was considered too simplistic and rejected by most physicists. While Messrs. Lorentz, Jeans and others made formidable objections to it, Mr. Einstein countered by showing how the study of fluctuations in black body radiation also led to the concept of a radiant energy discontinuity. The International Congress of Physics held in Brussels in 1911 under the auspices of Mr, Solvay was entirely devoted to the question of quanta and it was in the aftermath of this congress that Henri Poincaré published a series of papers on quanta shortly before his death, showing the need to accept Planck's ideas.

In 1913, Mr. Niels Bohr's atom theory was published. He admitted with Messrs. Rutherford and Van Den Broek that the atom is formed of a positive nucleus surrounded by a cloud of electrons, the nucleus carrying $N$ positive elementary charges $4.77 \times 10^{-10}$ e.u.s., and the number of electrons being $N$ so that the whole is neutral. $N$ is the atomic number equal to the order number of the element in the Mendeleev periodic series. In order to be able to predict the optical frequencies especially for hydrogen whose single electron atom is especially simple, Bohr makes two hypotheses:

1° Among the infinity of trajectories that an electron translating around the nucleus can describe, only some are stable and the condition of stability involves Planck's constant. We will specify in chapter III the nature of these conditions;

2° When an intraatomic electron passes from one stable trajectory to another, there is emission or absorption of a quantum of energy of frequency $\nu$. The emitted or absorbed frequency $\nu$ is thus related to the variation $\delta\epsilon$ of the total energy of the atom by the relation $|\delta\epsilon| = h\nu$.

We know what the magnificent success of Bohr's theory has been over the last ten years. It immediately allowed the prediction of the spectral series of hydrogen and ionized helium: the study of $X$-ray spectra and the famous Moseley's law linking the atomic number to the spectral landmarks of the Röntgen domain have considerably extended the scope of its application. Sommerfeld, Epstein, Schwarzschild, Bohr himself and others perfected the theory, stated more general quantification conditions, explained the Stark and Zeemann effects, interpreted the optical spectra in detail, etc. But the deeper meaning of quanta has remained unknown. The study of the photoelectric effect of $X$-rays by Mr. Maurice de Broglie, that of the photoelectric effect of $\gamma$-rays due to Messrs. Rutherford and Ellis have more and more accentuated the corpuscular character of these radiations, the energy quantum $h\nu$ seeming each day more and more to constitute a real atom of light. But the old objections against this view remained and, even in the field of $X$-rays, the wave theory was successful: prediction of Laue interference and scattering phenomena (works of Debye, W.-L. Bragg, etc.). However, very recently, scattering in its turn has been submitted to the corpuscular point of view by Mr. H.-A. Compton: his theoretical and experimental works have shown that an electron diffusing some radiation must suffer a certain impulse as in a shock; naturally the energy of the radiation quantum is diminished and, as a consequence, the scattered radiation has a frequency that varies according to the direction of scattering and is lower than the frequency of the incident radiation. In short, the time seemed ripe to attempt an effort to unify the corpuscular and wave points of view and to deepen further the true meaning of quanta. This is what we have done recently and the main purpose of this thesis is to present a more complete account of the new ideas we have proposed, the successes they have led us to, and also the very numerous shortcomings that they entail.[1]

[1]Let us mention here a few works that deal with quanta issues:

J. Perrin, *Les atomes*, Alcan, 1913.

H. Poincaré, *Dernières pensées*, Flammarion, 1913.

E. Bauer, Recherches sur le rayonnement, *Thèse de doctorat*, 1912.

*La théorie du rayonnement et les quanta* (1er Congrès Solvay, 1911), publiée par P. Langevin et M. de Broglie.

M Planck, *Theorie der Wärmestrahlung*, J.-A. Barth, Leipzig, 1921 (4e édit.).

L. Brilloin, *La théorie des quanta et l'atome de Bohr* (Conf. rapports), 1921.

F. Reiche, *Die quantentheorie*, J. Springer, Berlin, 1921.

A. Sommerfeld, *La constitution de l'atome et les rates spectrales*. Trad. Bellenot, A. Blanchard, 1923.

A. Landé, *Vorschritte der quantentheorie*, F. Steinhopff, Dresde, 1922. Atomes et électrons (3e Congrès Solvay), Gauthier-Villars, 1923.

# 1 The Phase Wave

## I. The Relation between Quantum and Relativity

One of the most important new concepts introduced by Relativity is that of the inertia of energy. According to Einstein, energy must be considered as having mass and all mass represents energy. Mass and energy are always related to each other by the general relationship:

$$\text{Energy} = \text{mass } c^2$$

$c$ being the constant called "speed of light" but which we prefer to call "limit speed of energy" for reasons explained later. Since there is always proportionality between mass and energy, we must consider matter and energy as two synonymous terms designating the same physical reality.

Atomic theory first, and then electronic theory have taught us to consider matter as essentially discontinuous, and this leads us to admit that all forms of energy, contrary to ancient ideas about light, are if not entirely concentrated in small portions of space, at least condensed around certain singular points.

The principle of the inertia of energy assigns to a body whose proper mass (i.e., measured by an observer related to it) is $m_0$ a proper energy $m_0c^2$. If the body is in uniform motion with a velocity $v = \beta c$ with respect to an observer that we will name, to simplify, the fixed observer, its mass will have for this one the value $\frac{m_0}{\sqrt{1-\beta^2}}$ according to a well-known result of Relativistic Dynamics and, consequently, its energy will be $\frac{m_0 c^2}{\sqrt{1-\beta^2}}$. Since kinetic energy can be defined as the increase that the energy of a body experiences for the fixed observer when it passes from rest to the speed $v = \beta c$, we find the following expression for its value:

$$E_{\text{kin}} = \frac{m_0 c^2}{\sqrt{1-\beta^2}} - m_0 c^2 = m_0 c^2 \left( \frac{1}{\sqrt{1-\beta^2}} - 1 \right),$$

which naturally for the small values of $\beta$ leads to the classical form:

$$E_{\text{kin}} = \frac{1}{2} m_0 v^2.$$

This being posed, let us investigate under which form quanta can be involved in the dynamics of Relativity. It seems that the fundamental idea of quantum theory is the impossibility of considering an isolated energy quantum without relating some frequency to it. This relation is expressed by what I will name the quantum relationship:

$$\text{energy} = h \times \text{frequency}$$

$h$ being Planck's constant.

The progressive development of quantum theory has often focused on mechanical action, and numerous attempts have been made to define the quantum relationship as a statement involving action rather than energy. Certainly, the constant $h$ has the dimensions of an action, i.e. $ML^2T^{-1}$, and this is not due to chance since Relativity theory leads us to classify the action among the most important "invariants" of Physics. But action is a magnitude of a very abstract character and, after long pondering about light quanta and the photoelectric effect, we have been brought back to adopting the energy statement as a basis, even if this means investigating why action plays such an important role with regard to so many issues. The quantum relationship would probably not make much sense if energy could be distributed continuously in space, but we have just seen that this is probably not the case. It is therefore conceivable that as a result of a great law of Nature, to each amount of energy of proper mass $m_0$, is linked a periodic phenomenon of frequency $\nu_0$ such that we have:

$$h\nu_0 = m_0 c^2$$

$\nu_0$ being measured, of course, in the system linked to this amount of energy. This hypothesis is the foundation of our system: it is worth, like all hypotheses, what the consequences that can be deduced from it are worth.

Are we to assume that the periodic phenomenon is localized *within* the lump of energy? There is no need to do so and it will result from paragraph III that it is probably widespread over a large portion of space. Moreover, what should we mean by the interior of a lump of energy? The electron is for us the type of the isolated lump of energy which we believe, perhaps wrongly, to know best; but according to the received conceptions, the energy of the electron is spread throughout space with a very strong condensation in a region of very small dimensions whose properties are besides very badly known to us. What characterizes the electron as an atom of energy is not the small place that it occupies in space, I repeat that it occupies the whole space, it is the fact that it is unbreakable, not subdivisible, that it forms *a unit.*[1]

Having admitted the existence of a frequency related to the lump of energy, let's find out how this frequency manifests itself to the fixed observer mentioned above. The Lorentz-Einstein's transformation of time teaches us that a periodic phenomenon related to the moving body appears slowed down to the fixed observer in the ratio of 1 to $\sqrt{1-\beta^2}$. This is the famous slowing

[1]About the difficulties that arise when several electrically charged centres interact, see Chapter IV below.

down of clocks. Thus the frequency observed by the fixed observer will be

$$\nu_1 = \nu_0\sqrt{1-\beta^2} = \frac{m_0c^2}{h}\sqrt{1-\beta^2}.$$

On the other hand, since the energy of the moving entity[2] for the same observer is equal to $\frac{m_0c^2}{\sqrt{1-\beta^2}}$ the corresponding frequency according to the quantum relation is $\nu = \frac{1}{h}\frac{m_0c^2}{\sqrt{1-\beta^2}}$. The two frequencies $\nu_1$ and $\nu$ are essentially different since the factor $\sqrt{1-\beta^2}$ is not shown in the same way. There is here a difficulty that intrigued me for a long time; I managed to overcome it by demonstrating the following theorem, which I will call the phases harmony theorem:

> The periodic phenomenon related to the moving entity and whose frequency is for the fixed observer equal to $\nu_1 = \frac{1}{h}m_0c^2\sqrt{1-\beta^2}$ appears to the observer constantly in phase with a wave of frequency $\nu = \frac{1}{h}m_0c^2\frac{1}{\sqrt{1-\beta^2}}$ propagating in the same direction as the moving entity with velocity $V = c/\beta$.

The demonstration is very simple. Suppose that at time $t = 0$, there is phase agreement between the periodic phenomenon related to the moving entity and the wave defined above. At time $t$, the moving entity has crossed from the origin instant a distance equal to $x = \beta ct$ and the phase of the periodic phenomenon has varied by $\nu_1 t = \frac{m_0c^2}{h}\sqrt{1-\beta^2}\frac{x}{\beta c}$. The phase of the

[2] EDITOR'S NOTE: de Broglie used the term "mobile" but its meaning in English – "any body that is in motion" – cannot be used since it implies an object with classical behavior and may lead to confusion. That is why the term "moving entity" is used to represent de Broglie's "mobile."

portion of the wave that overlaps the moving body has varied by:

$$\nu\left(t - \frac{\beta x}{c}\right) = \frac{m_0 c^2}{h} \frac{1}{\sqrt{1-\beta^2}}\left(\frac{x}{\beta c} - \frac{\beta x}{c}\right)$$
$$= \frac{m_0 c^2}{h}\sqrt{1-\beta^2}\,\frac{x}{\beta c}.$$

As anticipated, the phases agreement persists.

It is possible to give another demonstration of this theorem which is basically the same, but perhaps more striking. If $t_0$ represents the time for an observer related to the moving entity (the proper time of the moving entity), the Lorentz transformation gives:

$$t_0 = \frac{1}{\sqrt{1-\beta^2}}\left(t - \frac{\beta x}{c}\right).$$

The periodic phenomenon that we imagine is represented for the same observer by a sinusoidal function of $\nu_0 t_0$. For the fixed observer, it is represented by the same sinusoidal function of $\nu_0 \frac{1}{\sqrt{1-\beta^2}}\left(t - \frac{\beta x}{c}\right)$ function which represents a wave of frequency $\frac{\nu_0}{\sqrt{1-\beta^2}}$ propagating with velocity $c/\beta$ in the same direction as the moving entity.

It is now essential to ponder about the nature of the wave whose existence we have just conceived. The fact that its velocity $V = \frac{c}{\beta}$ is necessarily greater than $c$ ($\beta$ being always less than 1, otherwise the mass would be infinite or imaginary), shows us that there could be no question of a wave transporting energy. Our theorem also shows us that it represents the spatial distribution of the *phases* of a phenomenon; it is a "phase wave".

To clearly explain this last point, we are going to propose a mechanical comparison which is a bit crude, but that speaks to the imagination. Let us suppose a circular horizontal plate of very large radius; from this plate are suspended identical systems consisting of a spiral spring to which a weigh is attached. The number of systems suspended in this way per unit area of the tray, their density, decreases very rapidly as one moves away from the centre of the plate so that there is a concentration of systems about the centre. All the spring-weight systems being identical, all have the same period; let us cause them to oscillate with the same amplitude and the same phase. The surface passing through the centres of gravity of all the weights will be a plane that will move up and down in a reciprocating motion. The assembly thus obtained presents a very rough analogy with an isolated lump of energy as we conceive it.

The description just provided is suitable for an observer attached to the set. If another observer sees the plate moving with a uniform translational motion with velocity $v = \beta c$, each weight will appear to him as a small clock undergoing Einstein's slowing down; moreover, the plate and the distribution of the oscillating systems will no longer be isotropic around the center because of the Lorentz contraction. But the fundamental fact for us (the 3rd paragraph will be more explicit) is the phase shift of the motions of the different weights. If, at a given moment of his time, our stationary observer considers the geometrical location of the centres of gravity of the various weights, he obtains a cylindrical surface in the horizontal direction whose vertical sections parallel to the velocity of the plate are sinusoids. It corresponds in the particular case considered to our phase wave; according to the general theorem, this surface is animated with a velocity $\frac{c}{\beta}$ parallel to that of the plate and the frequency of vibration of a point of fixed abscissa that rests constantly on it is equal to the natural frequency of oscillation of the springs multiplied by $\frac{1}{\sqrt{1-\beta^2}}$. We can clearly see in this example (and this is our excuse for having insisted on it so long) how the phase wave corresponds to the transport of the phase and by no means to that of the energy.

The preceding results seem to us to be of extreme importance because, using a hypothesis strongly suggested by the very notion of quantum, they establish a link between the motion of a moving entity and the propagation of a wave and thus suggest the possibility of a synthesis of antagonistic theories on the nature of radiation. Already, we can note that the rectilinear propagation of the phase wave is linked to the rectilinear motion of the moving entity; the Fermat principle applied to the phase wave determines the shape of these rays which are straight lines while the Maupertuis principle applied to the moving entity determines its rectilinear trajectory which is one of the rays of the wave. In chapter II, we will try to generalize this mutual coincidence.

## II. Phase Velocity and Group Velocity

We now need to demonstrate an important relationship between the velocity of the moving entity and that of the phase wave. If waves of very similar frequencies propagate in the same direction $Ox$ with velocities $V$ that we will call phase propagation velocities, by their superposition, these waves will give rise to beat phenomena if velocity $V$ varies with frequency $\nu$. These phenomena have been studied in particular by Lord Rayleigh in the case of dispersive media.

Let us consider two waves of neighboring frequencies $\nu$ and $\nu' + \delta\nu$ and of velocities $V$ and $V' = V + (dV/d\nu)\delta\nu$; their superposition is analytically translated by the following equation obtained by neglecting the second number $\delta\nu$ in front of $\nu$:

$$\sin 2\pi\left(\nu t - \frac{\nu x}{V} + \phi\right) + \sin 2\pi\left(\nu' t - \frac{\nu' x}{V'} + \phi'\right)$$
$$= 2\sin 2\pi\left(\nu t - \frac{\nu x}{V} + \psi\right)\cos 2\pi\left[\frac{\delta\nu}{2}t - x\frac{d\left(\frac{\nu}{V}\right)}{d\nu}\frac{\delta\nu}{2} + \psi'\right].$$

So we have a resulting sine wave whose amplitude is modulated at frequency $\delta\nu$ because the sign of the cosine does not matter. This is a well-known result. If we designate by $U$ the propagation speed of the beat, or speed of the group of waves, we find:

$$\frac{1}{U}=\frac{d\left(\frac{\nu}{V}\right)}{d\nu}.$$

Let's come back to phase waves. If we assign to the moving entity a velocity $v=\beta c$ and do not give $\beta$ a specific value, but only require it to be between $\beta$ and $\beta+\delta\beta$; the frequencies of the corresponding waves will fill a small interval $\nu+\delta\nu$.

We will establish the following theorem that will be useful later. "The speed of the group of phase waves is equal to the speed of the moving entity". Indeed, this group velocity is determined by the formula given above in which $V$ and $\nu$ can be considered being a function of $\beta$, since we have:

$$V=\frac{c}{\beta},\qquad \nu=\frac{1}{h}\frac{m_0c^2}{\sqrt{1-\beta^2}}.$$

We can write:

$$U=\frac{\frac{d\nu}{d\beta}}{\frac{d\left(\frac{\nu}{V}\right)}{d\beta}}$$

But:

$$\frac{d\nu}{d\beta}=\frac{m_0c^2}{h}\frac{\beta}{(1-\beta^2)^{3/2}}$$

$$d\frac{\left(\frac{\nu}{V}\right)}{d\beta}=\frac{m_0c}{h}\frac{d\left(\frac{\beta}{\sqrt{1-\beta^2}}\right)}{d\beta}=\frac{m_0c}{h}\frac{1}{(1-\beta^2)^{3/2}}.$$

Then:

$$U=\beta c=v.$$

The group velocity of the phase waves is effectively equal to the velocity of the moving entity. This result calls for a remark: in the wave theory of dispersion, if we exclude the absorption areas, the velocity of the energy is equal to the group velocity.[3]

[3]See for example Léon Brillouin. *La theéorie des quanta et l'atome de Bohr*, Chapter I.

Here, although considered from a very different point of view, we find a similar result, because the velocity of the moving body is nothing other than the displacement velocity of the energy.

## III. The Phase Wave in Space-Time

Minkowski was the first to show that a simple geometrical representation of the relations of space and time introduced by Einstein was obtained by considering a 4-dimensional Euclidean multiplicity[4] called Universe or Space-time.[5] To do this he took 3 axes of rectangular space coordinates and a fourth axis normal to the first 3 on which the times multiplied by $c\sqrt{-1}$ were carried. Today, the real quantity $ct$ is more readily related to the fourth axis, but then the planes passing through this axis are normal to space and have a hyperbolic pseudo-Euclidean geometry whose fundamental invariant is $c^2dt^2 - dx^2 - dy^2 - dz^2$.

Let us thus consider space-time related to the 4 orthogonal axes of the observer known as "stationary". We will take as the $x$-axis the rectilinear trajectory of the moving entity and we will represent on our sheet of paper the plane $otx$ containing the time axis and the said trajectory. Under these conditions, the worldline[6] of the moving entity is represented by a straight line tilted less than 45° on the time axis; this line is the time axis for the observer linked to the moving entity. We represent on our figure the 2 time axes intersecting at the origin, which does not restrict the generality.

If the speed of the moving entity for the stationary observer

---

[4] EDITOR'S NOTE: The translators kept de Broglie's word "multiplicité." Now the accepted word is "manifold" or simply "space."

[5] EDITOR'S NOTE: Here de Broglie did not mention Poincaré's 1905 paper, where he noticed that the Lorentz transformations can be viewed as rotations in a four-dimensional space – see EDITOR'S NOTE at the end of this Chapter.

[6] EDITOR'S NOTE: De Broglie used the expression "la line d'Univers;" now the accepted expression is "worldline" which is a direct translation of the novel term introduced by Minkowski "Weltlinie."

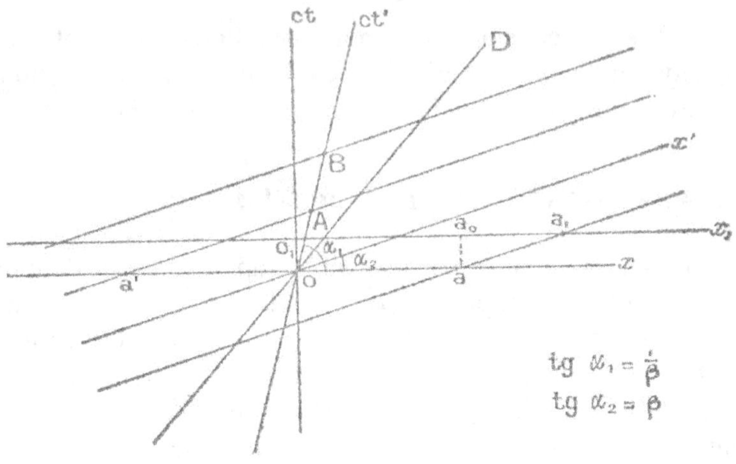


Fig. 1

is $\beta c$, the slope of $Ot'$ has the value $1/\beta$. The line $ox'$, plotted on the $tox$ plane of the space of the moving observer at time $O$, is symmetrical of $Ot'$ with respect to bisector $OD$; it is easy to demonstrate this analytically by means of the Lorentz transformation, but this results immediately from the fact that the velocity limit of energy $c$ has the same value for all reference systems. The slope of $Ox'$ is therefore $\beta$. If the space surrounding the moving entity is the seat of a periodic phenomenon, the state of the space will become the same for the moving observer each time a period of time $\frac{1}{c}\overline{OA} = \frac{1}{c}\overline{AB}$ equal to the proper period $T_0 = \frac{1}{\nu_0} = \frac{h}{m_0 c^2}$ of the phenomenon has elapsed.

The lines parallel to $ox'$ are thus the traces of these "equiphase spaces" of the observer moving on the $xot$ plane. The points... $a', o, a\ldots$ represent in projection their intersections with the space of the stationary observer at instant $O$; these intersections of 2 three-dimensional spaces are surfaces in 2 dimensions, and even planes, because all the spaces considered here are Euclidean. When time elapses for the stationary observer, the section of space-time which, for him, is space, is represented by a line parallel to $ox$ moving with a uniform motion towards the increasing $t$. One can easily see that the equiphase planes ... $a', o, a$ ... move in the space of the stationary observer with a velocity $\frac{c}{\beta}$.[7] Indeed, if line $ox_1$ of the figure represents the space of the stationary observer at time $t = 1$, one has $\overline{aa_0} = c$. The phase which for $t = O$, was in $a$, is now in $a_1$; for the stationary observer, it thus moved in its space by the length $a_0 a_1$ in direction $ox$ during the unit of time. We can thus say that its velocity is:

$$V = a_0 a_1 = aa_0 \operatorname{cotg}(\widehat{xox'}) = \frac{c}{\beta}.$$

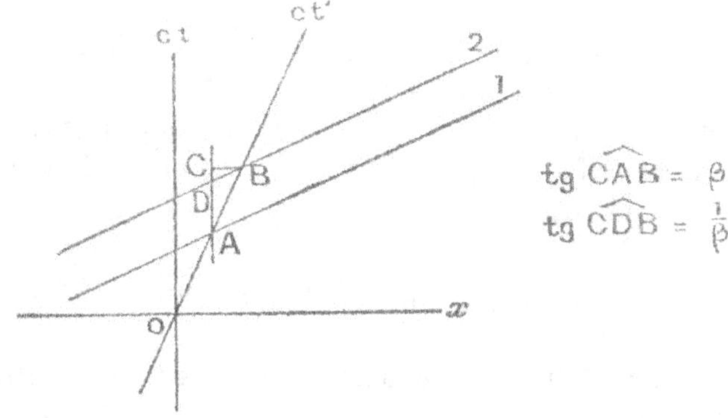


Fig. 2

The set of equiphase planes constitutes what we have named the phase wave. The issue of frequencies remains to be addressed. Let's redo a small simplified figure.

Lines 1 and 2 represent two successive equidistant spaces of the linked observer. $\overline{AB}$ is, as previously mentioned, equal to $c$ times the proper period $T_0 = \frac{h}{m_0 c^2}$.

$AC$, projection of $AB$ on axis $Ot$, is equal to

---

[7] EDITOR'S NOTE: This paragraph contains the very essence of de Broglie's novel idea, which, as clearly seen here, stems from Minkowski's explanation that *observers in relative motion have different spaces and times* (which led him to the realization that that is possible in an absolute four-dimensional world).

$$cT_1 = cT_0 = cT_0 \frac{1}{\sqrt{1-\beta^2}}.$$

This is the result of a simple application of trigonometric relations; however, we note that when applying trigonometry to figures on the $xot$ plane, one must always keep in mind the particular anisotropy of this plane. The triangle $ABC$ gives us:

$$\begin{aligned}\overline{AB}^2 &= \overline{AC}^2 - \overline{CB}^2 = \overline{AC}^2(1 - \mathrm{tg}^2\,\widehat{\mathrm{CAB}})\\ &= \overline{AC}^2(1-\beta^2)\\ \overline{AC} &= \frac{\overline{AB}}{\sqrt{1-\beta^2}} \qquad\qquad \text{c. q. f. d.}\end{aligned}$$

Frequency $1/T_1$ is the frequency that the periodic phenomenon appears to have for the stationary observer who follows it with his eyes as it moves. It is:

$$\nu_1 = \nu_0\sqrt{1-\beta^2} = \frac{m_0c^2}{h}\sqrt{1-\beta^2}.$$

The period of the waves at a point in space for the stationary observer is given not by $\frac{1}{c}\overline{AC}$, but by $\frac{1}{c}\overline{AD}$. Let's calculate it.

In the small $BCD$ triangle, we find the relation

$$\frac{\overline{CB}}{\overline{DC}} = \frac{1}{\beta} \qquad \text{or} \qquad \overline{DC} = \beta\overline{CB} = \beta^2\overline{AC}.$$

But $\overline{AD} = \overline{AC} - \overline{DC} = \overline{AC}(1-\beta^2)$. The new period $T$ is therefore equal to:

$$T = \frac{1}{c}\overline{AC}(1-\beta^2) = T_0\sqrt{1-\beta^2}$$

and the frequency $\nu$ of the waves is expressed by:

$$\nu = \frac{1}{T} = \frac{\nu_0}{\sqrt{1-\beta^2}} = \frac{m_0c^2}{h\sqrt{1-\beta^2}}.$$

We thus recover all the results obtained analytically in the first paragraph, but now we see better how they relate to the general conception of space-time and why the phase shift of periodic motions taking place at different points in space depends on the manner in which simultaneity is defined by the theory of Relativity.

## Editor's Note

on the following assertion by de Broglie on p. 19:

> Minkowski was the first to show that a simple geometrical representation of the relations of space and time introduced by Einstein was obtained by considering a 4-dimensional Euclidean multiplicity called Universe or Space-time.

It is remarkable that a French physicist gave Hermann Minkowski full credit for the discovery of the spacetime structure of the world without even mentioning Poincaré who first published his observation that the Lorentz transformations could be viewed as rotations in a four-dimensional space with time as the fourth dimension [1]. Although de Broglie could have added a footnote to acknowledge this fact, he was right to give the full credit to Minkowski because Poincaré noticed the geometrical interpretation only of Lorentz' transformations, whereas, to repeat de Broglie's precise statement, "Minkowski was the first to show that a simple geometrical representation of *the relations of space and time* introduced by Einstein was obtained by considering a 4-dimensional Euclidean multiplicity called Universe or Space-time" (italics added).

More importantly, Poincaré believed that time can only formally be regarded as a dimension of a four-dimensional space, which is nothing more than a *convenient* mathematical space; as another French physicist pointed out [1]:

> Although the first discovery of the mathematical structure of the space-time of special relativity is due to Poincaré's great article of July 1905, Poincaré (in contrast to Minkowski) had never believed that this structure could really be important for physics.

The first part of the quote needs clarification – Poincaré only noticed that the Lorentz transformations can be viewed as rotations in a four-dimensional space, but it was Minkowski who first and fully developed the four-dimensional formalism of space-time physics.[8]

Poincaré believed that our physical theories are only convenient descriptions of the world and therefore it is really a matter of convenience and our choice which theory we would use. As Damour stressed it [3], it was

> ... the sterility of Poincaré's scientific philosophy: complete and utter "conventionality" ... which stopped him from taking seriously, and developing as a physicist, the space-time structure which he was the first to discover.

What makes Poincaré's failure to comprehend the profound *physical* meaning of the relativity principle and the geometric interpretation of the Lorentz transformations especially sad is that it is perhaps the most cruel example in the history of physics of how an inadequate view on the nature of physical theories can prevent a scientist, even as great as Poincaré, from making a discovery.

1. H. Poincaré, "Sur la dynamique de l'électron," *Rendiconti del Circolo matematico Rendiconti del Circolo di Palermo* **21** (1906) pp. 129-176

2 . T. Damour, *Once Upon Einstein,* Translated by E. Novak (A. K. Peters, Wellesley 2006), p. 52

[8]Moreover, at least two things (one of which is Born's recollections) appear to indicate that Minkowski arrived independently at what Einstein called special relativity and at the concept of spacetime, but Einstein and Poincaré published first while Minkowski had been developing the four-dimensional formalism of spacetime physics published in 1908 as a 59-page treatise (*The Fundamental Equations for Electromagnetic Processes in Moving Bodies*; H. Minkowski, Die Grundgleichungen für die elektromagnetischen Vorgänge in bewegten Körpern, *Nachrichten der K. Gesellschaft der Wissenschaften zu Göttingen. Mathematisch-physikalische Klasse* (1908) S. 53-111) – for details see `http://www.minkowskiinstitute.org/born.html`.

# 2 PRINCIPLE OF MAUPERTUIS AND PRINCIPLE OF FERMAT

## I. Purpose of this Chapter

In this chapter we will try to generalize the results of chapter one for the case of a moving entity whose motion is not rectilinear and uniform. Non-uniform motion presupposes the existence of a force field to which the moving object is subjected. In the current state of our knowledge there seems to be only two sorts of fields: gravitational fields and electromagnetic fields. The theory of General Relativity[1] interprets the gravitational field as due to a curvature of space-time. In the present thesis, we will systematically leave aside everything concerning gravitation, eventually coming back to it in another work. For us, therefore, at this point in time, a force field will be an electromagnetic field and the dynamics of the non-uniform motion will be the study of motion of a body carrying an electric charge in an electromagnetic field.

We must expect to meet in this chapter some great difficulties because the Relativity Theory, which is a very sure guide when it comes to uniform motion, is still quite hesitant in its conclusions about non-uniform motion.[2] During Einstein's re-

[1] EDITOR'S NOTE: de Broglie used the expression "The theory of Generalized Relativity" (La théory de Relativité généralisée).

[2] EDITOR'S NOTE: Only what Einstein called special relativity might be hes-

cent stay in Paris, Mr. Painlevé raised some amusing objections against Relativity; Mr. Langevin was able to dismiss them without difficulty since they all involved accelerations whereas the Lorentz-Einstein transformation applies only to uniform motion. However, the arguments of the illustrious mathematician proved once again that the application of Einstein's ideas becomes very delicate when dealing with accelerations and, in this respect, they are very instructive. The method that allowed us to study the phase wave in Chapter 1 will no longer be of any help here.

The phase wave that accompanies the motion of a moving entity, if we admit our concepts, has properties that depend on the nature of this moving entity since its frequency, for example, is determined by its total energy. It therefore seems natural to assume that if a force field acts on the motion of a moving object, it will also act on the propagation of its phase wave. Guided by the idea of a profound identity between the principle of least action and that of Fermat, I was led from the beginning of my research on this subject to admit that for a given value of the total energy of the moving entity and as a consequence of the frequency of its phase wave, the dynamically possible trajectories of one coincided with the possible rays of the other. This led me to a very satisfactory result which will be presented in Chapter III, namely the interpretation of the conditions of intra-atomic stability established by Bohr. Unfortunately, fairly arbitrary assumptions were needed regarding the value of the propagation velocities $V$ of the phase wave at each point of the field. We will, on the contrary, use here a method that seems much more general and more satisfactory. We will study on the one hand the mechanical principle of least action in its Hamiltonian and Maupertuisian forms in classical dynamics and in Relativity and on the other hand, from a very general point of view, wave propagation and

---

itant about the status of acceleration, whereas in Minkowski 's spacetime presentation in 1908 "the concept of *acceleration* acquires a sharply prominent character" – see EDITOR'S NOTE 1 on acceleration in the theory of relativity at the end of this Chapter.

the Fermat principle. We will then be led to conceive a synthesis of these two studies, a synthesis which can be debatable but whose theoretical elegance is indisputable. At the same time, we will obtain the solution to the problem posed.

## II. The Two Principles of Least Action in Classical Dynamics

In classical dynamics, the principle of least action in its Hamiltonian form is stated as follows: "The equations of dynamics can be deduced from the fact that the integral

$$\int_{t_1}^{t_2} \mathfrak{L} dt$$

taken between fixed time limits for initial and final values given by parameters $q_i$ that determine the state of the system, has a stationary value."

By definition, $\mathfrak{L}$ is named the Lagrange function and is supposed to depend on variables $q_i$ and $\dot{q}_i = dq_i/dt$.

So we have:

$$\delta \int_{t_1}^{t_2} \mathfrak{L} dt = 0.$$

This is deduced by a known method of calculating variations, the so-called Lagrange equations:

$$\frac{d}{dt}\left(\frac{\partial \mathfrak{L}}{\partial \dot{q}_i}\right) = \frac{\partial \mathfrak{L}}{\partial q_i}.$$

in equal number to the number of variables $q_i$.

Function $\mathfrak{L}$. remains to be defined. The classical dynamics poses:

$$\mathfrak{L} = E_{\text{kin}} - E_{\text{pot}}$$

or the difference between kinetic and potential energies. We will see further on that relativistic dynamics uses a different value for $\mathfrak{L}$.

Let us now move on to the Maupertuisian form of the principle of least action. For this, let us first note that the Lagrange equations in the general form given above admit a first integral termed "energy of the system" equal to:

$$W = -\mathfrak{L} + \sum_i \frac{\partial \mathfrak{L}}{\partial \dot{q}_i} \dot{q}_i$$

provided, however, that function $\mathfrak{L}$ is not explicitly time-dependent, which we will always assume hereafter. We have indeed then:

$$\begin{aligned}\frac{dW}{dt} &= -\sum_i \frac{\partial \mathfrak{L}}{\partial q_i} \dot{q}_i - \sum_i \frac{\partial \mathfrak{L}}{\partial \dot{q}_i} \ddot{q}_i + \sum_i \frac{\partial \mathfrak{L}}{\partial q_i} \ddot{q}_i + \sum_i \frac{d}{dt}\left(\frac{\partial \mathfrak{L}}{\partial \dot{q}_i}\right) \dot{q}_i \\ &= \sum_i \dot{q}_i \left[\frac{d}{dt}\left(\frac{\partial \mathfrak{L}}{\partial \dot{q}_i}\right) - \frac{\partial \mathfrak{L}}{\partial q_i}\right],\end{aligned}$$

which is a null quantity according to the Lagrange equations. Therefore, we can say:

$$W = C^{te}.$$

Let us now apply the Hamiltonian principle to all of the "various" trajectories that lead from a given initial state $A$ to a given final state $B$ and that correspond to a given value of the energy W. We can write since $W$, $t_1$ and $t_2$ are constant:

$$\delta \int_{t_1}^{t_2} \mathfrak{L} dt = \delta \int_{t_1}^{t_2} (\mathfrak{L} + W) dt = 0$$

or even better:

$$\delta \int_{t_1}^{t_2} \sum_i \frac{\partial \mathfrak{L}}{\partial \dot{q}_i} dt = \delta \int_A^B \sum_i \frac{\partial \mathfrak{L}}{\partial \dot{q}_i} dq_i = 0$$

the last integral being extended to all the $q_i$ values between those defining states $A$ and $B$ so that time is eliminated; there is therefore no longer any need in the new form obtained to impose any restrictions relative to time limits. On the other hand, the various trajectories must all correspond to the same energy value $W$.

Let us assume the classical notation of canonical equations: $p_i = \frac{\partial \mathfrak{L}}{\partial \dot{q}_i}$. The $pi$ are the conjugated moments of the variables $q_i$. The Maupertuisian principle is written:

$$\delta \int_A^B \sum_i p_i dq_i = 0$$

in classical dynamics where $\mathfrak{L} = E_{\text{kin}} - E_{\text{pot}}$, $E_{\text{pot}}$ is independent of $q_i$ and $E_{\text{kin}}$ is a homogeneous quadratic function. According to Euler's theorem:

$$\sum_i p_i dq_i i = \sum_i p_i \dot{q}_i dt = 2E_{\text{kin}} dt.$$

For the material point, $E_{\text{kin}} = \frac{1}{2} m v^2$ and the principle of least action takes its oldest known form:

$$\delta \int_A^B m v dl = 0$$

## III. The Two Principles of Least Action in the Dynamics of the Electron

We will now take up the question again for the dynamics of the electron from the relativistic point of view. The word "electron" must be taken here in the general sense of a material point carrying an electric charge. We will assume that the electron placed outside any field has a proper mass $m_0$; its electric charge is designated by *e*.

We will again consider space-time; the space coordinates will be named $x^1, x^2$ and $x^3$, the $ct$ coordinate will be $x^4$. The invariant fundamental "element of length" is defined by:

$$ds = \sqrt{(dx^4)^2 - (dx^1)^2 - (dx^2)^2 - (dx^3)^2}.$$

In this paragraph and in the following one, we will constantly use certain notations of tensor calculus.

A worldline has at each point a tangent whose direction is defined by the "four velocity" vector of unit length whose contravariant components are given by the relation:

$$u^i = \frac{dx^i}{ds} \qquad (i = 1,2,3,4).$$

We immediately verify that we have: $u^i u_i = 1$.

Either a moving entity travelling along the world line; when it passes the point considered, it has a velocity $v = \beta c$ with components $v_x, v_y, v_z$. The components of the four-velocity are:

$$u_1 = -u^1 = -\frac{v_x}{c\sqrt{1-\beta^2}} \qquad u_2 = -u^2 = -\frac{v_y}{c\sqrt{1-\beta^2}}$$

$$u_3 = -u^3 = -\frac{v_z}{c\sqrt{1-\beta^2}} \qquad u_4 = u^4 = \frac{1}{\sqrt{1-\beta^2}}$$

To define an electromagnetic field, we must introduce a second four-vector whose components are expressed as a function of the vector potential $\vec{a}$ and of the scalar potential $\Psi$ by the relations:

$$\phi_1 = -\phi^1 = -a_x; \qquad \phi_2 = -\phi^2 = -a_y; \qquad \phi_3 = -\phi^3 = -a_z$$

$$\phi_4 = \phi^4 = \frac{1}{c}\psi.$$

Let us now consider two points $P$ and $Q$ in space-time corresponding to given values of space and time coordinates. We can consider a curvilinear integral taken along a worldline from $P$ to $Q$; of course the function to be integrated must be invariant. Let:

$$\int_P^Q (-m_0 c - e\phi_i u^i) ds = \int_P^Q (-m_0 c u_i - e\phi_i) u^i ds$$

be this integral. Hamilton's principle states that if the worldline of a moving entity passes through $P$ and $Q$, it has a shape such that the integral defined above has a stationary value. Let's define a third four-vector by the relation:

$$\mathbf{J}_i = m_0 c u_i + e\phi_i \qquad (i = 1,2,3,4).$$

A little further on, we will give a physical meaning to four-vector **J**.

For the moment, let us return to the usual form of dynamic equations by replacing in the first form the action integral $ds$ by $cdt\sqrt{1-\beta^2}$. This gives us the following result:

$$\delta\int_{t_1}^{t_2}\left[-m_0c^2\sqrt{1-\beta^2} - ec\phi_4 - e\left(\phi_1 v_x + \phi_2 v_y + \phi_3 v_z\right)\right]dt = 0$$

here $t_1$ and $t_2$ correspond to points $P$ and $Q$ in space-time.

If there is a purely electrostatic field, the $\phi_1, \phi_2, \phi_3$ quantities are zero and the Lagrange function takes the form often used:

$$\mathfrak{L} = -m_0c^2\sqrt{1-\beta^2} - e\psi.$$

In all cases, since the Hamilton principle always has the form

$$\delta\int_{t_1}^{t_2}\mathfrak{L}dt = 0$$

one is always led to the Lagrange equations:

$$\frac{d}{dt}\left(\frac{\partial\mathfrak{L}}{\partial\dot{q}_i}\right) = \frac{\partial\mathfrak{L}}{\partial q_i} \qquad (i = 1,2,3).$$

In all cases where potentials are not time-dependent we find energy conservation:

$$W = -\mathfrak{L} + \sum_i p_i q_i = C^{te} \qquad p_i = \frac{\partial\mathfrak{L}}{\partial\dot{q}_i} \qquad i = 1,2,3.$$

By following exactly the same method as above, we obtain the Maupertuis principle:

$$\delta\int_A^{B_\beta}\sum_i p_i dq_i = 0,$$

where $A$ and $B$ are the two points in space that correspond to the reference system used at points $P$ and $Q$ in space-time.

The quantities $p_1, p_2, p_3$ which are equal to the partial derivatives of the function $\mathfrak{L}$ with respect to the corresponding velocities can be used to define a vector $\vec{p}$ which we will call the "momentum vector." If there is no magnetic field (whether or not there is an electric field), the rectangular components of this vector are:

$$p_x = \frac{m_0 v_x}{\sqrt{1-\beta^2}} \qquad p_y = \frac{m_0 v_y}{\sqrt{1-\beta^2}} \qquad p_z = \frac{m_0 v_z}{\sqrt{1-\beta^2}}.$$

It is therefore identical to the momentum and the Maupertuisian action integral has the simple form proposed by Maupertuis himself with the only difference that mass now varies with velocity according to Lorentz' law.[3]

If there is a magnetic field, we find for the components of the momentum vector the expressions:

$$p_x = \frac{m_0 v_x}{\sqrt{1-\beta^2}} + e a_x$$
$$p_y = \frac{m_0 v_y}{\sqrt{1-\beta^2}} + e a_y$$
$$p_z = \frac{m_0 v_z}{\sqrt{1-\beta^2}} + e a_z.$$

There is no longer identity between the $\vec{p}$ vector and the momentum; as a result, the expression for the action integral becomes more complicated.

Let's consider a moving entity placed in a field and whose total energy is given; at any point in the field that the moving entity can reach, its velocity is given by the energy equation, but the $a$

[3] EDITOR'S NOTE: I cannot resist the temptation to add a note (see EDITOR'S NOTE 2 on relativistic mass at the end of this Chapter) because I suspect some over-confident colleagues (perhaps mostly particle physicists) would remark that de Broglie employed velocity-dependent mass, which they label an old-fashioned concept.

*priori* direction can be any direction. The expression of $p_x, p_y$ and $p_z$ shows that the momentum vector has the same magnitude at a point of an electrostatic field, whatever the direction considered. It is no longer the same if there is a magnetic field: the magnitude of vector $\vec{p}$ then depends on the angle between the chosen direction and the potential vector as seen by forming the expression $px^2 + py^2 + pz^2$. This remark will be useful later on.

To end this paragraph, we will come back to the physical meaning of world line vector **J** on which the Hamiltonian integral depends. We defined it by the expression:

$$J_i = m_0 c\, u_i + e\phi_i \qquad (i = 1,2,3,4).$$

By means of the values $u_i$ and $\phi_i$ we find:

$$J_1 = -p_x \qquad J_2 = -p_y \qquad J_3 = -p_z \qquad J_4 = \frac{W}{c}.$$

The contravariant components will be:

$$J^1 = p_x \qquad J^2 = p_y \qquad J^3 = p_z \qquad J^4 = \frac{W}{c}.$$

We are therefore dealing with the famous "four-momentum" vector[4] that synthesizes the energy and the momentum. From:

$$\delta \int_P^Q J_i dx^i = 0 \qquad (i = 1,2,3,4)$$

we can immediately draw, if $J_4$ is constant:

$$\delta \int_A^B J_i dx^i = 0 \qquad (i = 1,2,3).$$

This is the most concise manner to switch from one stationary action statement to the other.

[4] EDITOR'S NOTE: The term used by de Broglie is "Impulsion d'Univers."

## IV. Wave Propagation; Fermat Principle

We will study the propagation of the phase of a sinusoidal phenomenon by a method parallel to that of the last two paragraphs. For this purpose, we will place ourselves from a very general point of view and again we will have to consider space-time.

Let us consider the function $\sin\phi$ in which the differential is supposed to depend on space and time variables $x^i$. There are in space-time an infinity of world lines along which the function $\phi$ is constant.

The wave theory, as resulting in particular from the work of Huyghens and Fresnel, teaches us to distinguish among these lines some whose projections on the space of an observer are for him the "rays" in the usual sense of optics.

Let as previously $P$ and $Q$ be two points in space-time. If a world ray (un rayon d'Univers) passes through these two points, what is the law that will determine its shape?

We will consider the curvilinear integral $\int_P^Q d\phi$ and take as the principle determining the world ray, the statement of Hamiltonian form:

$$\delta \int_P^Q d\phi = 0.$$

The integral must, in fact, be stationary, otherwise disturbances leaving a certain point in space in phase concordance and intersecting at another point after having followed slightly different paths, would display different phases at that point.

The phase $\phi$ is an invariant; so if we pose:

$$d\phi = 2\pi\left(O_1 dx^1 + O_2 dx^2 + O_3 dx^3 + O_4 dx^4\right) = 2\pi O_i\, dx^i$$

the $O_i$ quantities, which are generally functions of the $x^i$, will be the covariant components of a four-vector, the Wave four-vector. If $l$ is the direction of the ray in the ordinary sense, we are usually led to consider for $d\phi$ the form:

$$d\phi = 2\pi\left(\nu dt - \frac{\nu}{V} dl\right)$$

$\nu$ is the frequency and $V$ the speed of propagation. We can pose then:

$$O_1 = -\frac{\nu}{V}\cos(x,l), \qquad O_2 = -\frac{\nu}{V}\cos(y,l),$$
$$O_3 = -\frac{\nu}{V}\cos(z,l), \qquad O_4 = \frac{\nu}{c}.$$

The Wave four-vector is thus decomposed into a time component proportional to the frequency and a *space* vector $\vec{n}$ in the direction of propagation and having a length of $\frac{\nu}{V}$. We will name it the "wave number" vector because it is equal to the inverse of the wavelength. If the frequency $\nu$ is constant, we are led to switch from the Hamiltonian form:

$$\delta \int_P^Q O_i dx^i = o$$

to the Maupertuisian form:

$$\delta \int_A^B O_1 dx^1 + O_2 dx^2 + O_3 dx^3 = 0,$$

where $A$ and $B$ are the points in space corresponding to $P$ and $Q$.

By replacing $O_1, O_2$ and $O_3$ by their values, we obtain:

$$\delta \int_A^B \frac{\nu dl}{V} = 0.$$

This Maupertuisian statement constitutes Fermat's principle. Just as in the previous paragraph it was sufficient to find the trajectory that a moving entity of given total energy passing through two given points, to know the distribution in the field of vector $\vec{p}$, in the same way here to find the ray of a wave of known frequency passing through two given points, it is sufficient to know the distribution in space of the wave number vector that determines at each point and for each direction the speed of propagation.

## V. Extension of the Quantum Relation

We have now reached the high point of this chapter. From the very beginning, we posed the following question: "When a moving entity moves with a non-uniform motion in a force field, how does its phase wave propagate?" Instead of trying, as I first did, to determine the propagation velocity at each point and in each direction, I will extend the quantum relation, somewhat hypothetical at this point, but whose deep agreement with the philosophy of the Relativity Theory is indisputable.

We were constantly led to pose $h\nu = \omega$, $\omega$ as being the total energy of the moving entity and $\nu$ as being the frequency of its phase wave. On the other hand, the previous paragraphs have taught us to define two four-vectors **J** and **O** which play perfectly symmetrical roles in the study of the motion of a moving entity and in that of the propagation of a wave.

Using these vectors, the relationship $h\nu = \omega$ becomes:

$$O_4 = \frac{1}{h} J_4.$$

The fact that two vectors have an equal component does not prove that the same is true for the others. However, by a very suitable generalization we will pose:

$$O_i = \frac{1}{h} J_i \qquad (i = 1, 2, 3, 4).$$

The variation $d\phi$ relative to an infinitely small portion of the phase wave has the value:

$$d\phi = 2\pi O_i dx^i = \frac{2\pi}{h} J_i dx^i.$$

We therefore arrive at the following statement

"The Fermat principle applied to the phase wave is identical to the Maupertuis principle applied to the moving entity; the dynamically possible trajectories of the moving entity are identical to the possible rays of the wave".

We believe that this idea of a deep relationship between the two main principles of Geometric Optics and Dynamics could be a valuable guide to achieve the synthesis of waves and quanta.

The hypothesis of the proportionality of both **J** and **O** vectors is a sort of extension of the quantum relation, the current statement of which is clearly insufficient since it involves energy, without mentioning its inseparable companion, the momentum. The new statement is much more satisfactory because it is expressed by the equality of two four-vectors.

## VI. Specific Cases; Discussions

The general concepts of the previous paragraph must now be applied to specific cases in order to clarify their meaning.

a) Let us first consider the rectilinear and uniform motion of a free moving entity. The hypotheses made at the beginning of Chapter 1 allowed us, thanks to the principle of Special Relativity, to make a complete study of this case. Let us see if we can find the predicted value for the propagation velocity of the phase wave:

$$V = \frac{c}{\beta}.$$

Here, we must pose:

$$\nu = \frac{W}{h} = \frac{m_0 c^2}{h\sqrt{1-\beta^2}},$$

$$\frac{1}{h}\sum_1^3 p_i\, dq_i = \frac{1}{h}\frac{m_0\beta^2 c^2}{\sqrt{1-\beta^2}}dt = \frac{1}{h}\frac{m_0\beta c}{\sqrt{1-\beta^2}}dl = \frac{\nu dl}{V},$$

hence $V = \frac{c}{\beta}$ We gave an interpretation of this result from the space-time point of view.

b) Let us consider an electron in an electrostatic field (the Bohr atom). We must assume that the phase wave has a frequency $\nu$ equal to the quotient per $h$ of the total energy of the moving entity, that is:

$$W = \frac{m_0 c^2}{\sqrt{1-\beta^2}} + e\psi = h\nu.$$

Since the magnetic field is null, we will simply have:

$$p_x = \frac{m_0 v_x}{\sqrt{1-\beta^2}}, \qquad \text{etc.,}$$

$$\frac{1}{h}\sum_1^3 p_i\, dq_i = \frac{1}{h}\frac{m_0 \beta c}{\sqrt{1-\beta^2}} dl = \frac{\nu}{V} dl,$$

from which:

$$V = \frac{\dfrac{m_0 c^2}{\sqrt{1-\beta^2}} + e\psi}{\dfrac{m_0 \beta c}{\sqrt{1-\beta^2}}} = \frac{c}{\beta}\left(1 + \frac{e\psi\sqrt{1-\beta^2}}{m_0 c^2}\right)$$

$$= \frac{c}{\beta}\left(1 + \frac{e\psi}{W - e\psi}\right) = \frac{c}{\beta}\frac{W}{W - e\psi}.$$

This result calls for several remarks. From a physical point of view, it means that the phase wave of frequenc $\nu = \frac{W}{h}$ propagates in the electrostatic field with a variable velocity from one point to another according to the value of the potential. The velocity $V$ depends indeed on $\psi$ directly by the term (generally small in front of the unit) $\frac{e\psi}{W - e\psi}$ and indirectly by $\beta$ which is calculated at each point according to $W$ and $\psi$.

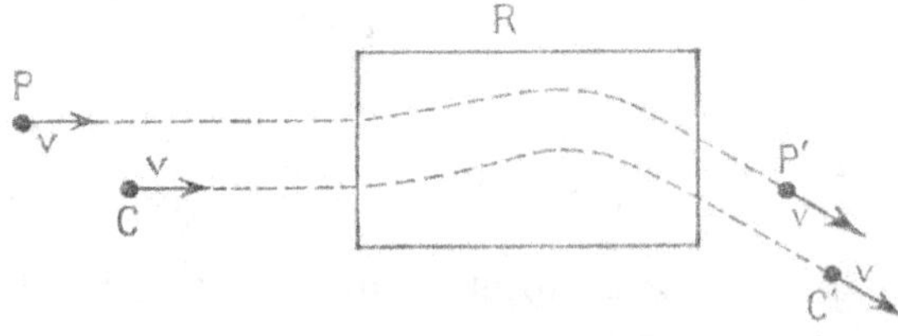


Fig. 3

Moreover, we will notice that $V$ is a function of the mass and the charge of the moving entity. This point may seem strange, but it is actually less odd than it seems. Let's consider an electron whose center $C$ moves with velocity $v$; in the classical conception, at a point $P$ whose coordinates in a system related to the electron are known, there is some electromagnetic energy that is somehow part of the electron. Suppose that after passing through a region $R$ where a more or less complex electromagnetic field prevails, the electron will be animated with the same velocity $v$ but in a different direction.

Point $P$ of the system linked to the electron went to $P'$ and we can say that the energy originally in $P$ was transported to $P'$. The displacement of this energy, even if we know the prevailing fields in $R$, can be calculated only if the mass and charge of the electron are given. This unquestionable conclusion could for a moment seem strange because we have the ingrained habit of considering mass and charge (as well as momentum and energy) as quantities related to the center of the electron. In the same way for the phase wave which, according to us, must be considered as a constituent part of the electron, the propagation in a field must depend on charge and mass.

Let us now recall the results obtained in the previous chapter in the case of uniform motion. We were then led to consider the phase wave as due to the intersections by the present space of the fixed observer, of the past, present and future spaces of the moving observer.[5] We could be tempted here again to find the above given value of $V$ by studying the successive "phases" of the moving entity and by specifying the displacement for the fixed observer of the sections by his space of the equiphase states. Un-

---

[5]EDITOR'S NOTE: In this sentence

> We were then led to *consider the phase wave as due to the intersections by the present space of the fixed observer, of the past, present and future spaces of the moving observer* (italics added).

de Broglie summarizes his novel idea based on Minkowski's realization that *observers in relative motion have different spaces and times* (which played a crucial role in Minkowski's path to the concept of spacetime); de Broglie's detailed explanation is on p. 20.

fortunately, we come up against very great difficulties here. Relativity does not currently teach us how a moving observer in non-uniform motion slices at each instant his space in space-time;[6] there does not seem to be much reason for this section to be flat as in uniform motion. But if this difficulty were solved, we would still be in trouble. Indeed, a moving object in uniform motion must be described in the same way by the observer related to it, whatever the velocity of the uniform motion, with respect to reference axes; this results from the principle that Galilean axes having uniform translational motions with respect to each other are equivalent. If therefore our moving entity in uniform motion is surrounded, for a related observer, by a periodic phenomenon having everywhere the same phase, it must be the same for all velocities of uniform motion and this is what justifies our method of chapter one. But if the motion is not uniform, the description of the moving entity made by the linked observer may not be the same and we do not know at all how he will define the periodic phenomenon and if he will attribute to it the same phase everywhere in space.

Perhaps one could reverse the problem, admit the results obtained in this chapter from quite different considerations and try to deduce how the Relativity theory must consider these questions of non-uniform motion in order to reach the same conclusions. We cannot address this difficult problem.

c) Let us take the general case of the electron in an electro-

[6] EDITOR'S NOTE: In its original 1905 formulation relativity did not teach us a lot of things. In Minkowski's 1908 spacetime approach there is no such difficulty – at any instant of an accelerated observer's proper time his space coincides with the space of an instantaneously comoving at that instant inertial observer; in other words, at a given event of the worldline of the accelerated observer his space coincides with the space of the instantaneously comoving at that event inertial observer, whose straight worldline (time axis) is tangent to the worldline of the accelerated observer at the event in question.

magnetic field. We still have:

$$h\nu = W = \frac{m_0 c^2}{\sqrt{1-\beta^2}} + e\psi.$$

In addition, we demonstrated previously that it was essential to pose:

$$p_x = \frac{m_0 v_x}{\sqrt{1-\beta^2}} + e a_x, \qquad \text{etc.,}$$

$a_x$, $a_y$ and $a_z$ are the components of the vector potential.

Therefore:

$$\frac{1}{h}\sum_1^3 p_i\, dq_i = \frac{1}{h}\frac{m_0 \beta c}{\sqrt{1-\beta^2}} dl + \frac{e}{h} a_l\, dl = \frac{\nu}{V} dl.$$

We thus find:

$$V = \frac{\dfrac{m_0 c^2}{\sqrt{1-\beta^2}} + e\psi}{\dfrac{m_0 \beta c}{\sqrt{1-\beta^2}} + e a_l} = \frac{c}{\beta}\frac{W}{W - e\psi}\frac{1}{1 + e\frac{a_l}{G}},$$

where $G$ is the amount of motion and $a_l$ the projection of the vector potential on direction $l$.

The medium at each point is no longer isotropic. Velocity $V$ varies with the direction being considered and the velocity of the moving entity $\vec{v}$ does not have the same direction as the normal to the phase wave defined by vector $\vec{p} = h\vec{n}$. The radius no longer coincides with the normal to the wave, which is the classical conclusion in the optics of anisotropic media.

One may wonder what happens to the equal velocity theorem $v = \beta c$ of the moving entity and the group velocity of the phase waves.

Let us first note that velocity $V$ of the phase that follows the ray is defined by relation:

$$\frac{1}{h}\sum_1^3 p_i\, dq^i = \frac{1}{h}\sum_1^3 p_i \frac{dq^i}{dl} dl = \frac{\nu}{V} dl,$$

in which $\frac{\nu}{V}$ is not equal to $\frac{1}{h}p$ because here $dl$ and $p$ do not have the same direction.

We can, without affecting generality, take as the $x$-axis the direction of motion of the moving entity at the point considered and name $p_x$ the projection of vector $\vec{p}$ on this direction. We then have the defining equation:

$$\frac{\nu}{V} = \frac{1}{h} p_x.$$

The first of the canonical equations provides equality:

$$\frac{dq}{dt} = v = \beta c = \frac{\partial W}{\partial p_x} = \frac{\partial (h\nu)}{\partial \left(h \frac{\nu}{V}\right)} = U,$$

where $U$ is the group speed according to the ray.

The result of Chapter 1, § 2, is therefore quite general and derives, in short, directly from the equations of Hamilton's first group.

## Editor's Note 1

on the following assertion by de Broglie on p. 27:

> We must expect to meet in this chapter some great difficulties because the Relativity Theory, which is a very sure guide when it comes to uniform motion, is still quite hesitant in its conclusions about non-uniform motion.

It is unfortunate that de Broglie did not elaborate more on this point because there had existed some misconceptions that special relativity did not deal with accelerated motion (apparently shared by Langevin and perhaps, to some extent, by de Broglie himself as judged from the sentence mentioning Langevin).

As de Broglie fully credited Minkowski for the discovery of the spacetime structure of the world (certainly due to a careful study[7] of Minkowski's papers on the subject), he should have been aware that in Minkowski's presentation of what Einstein called special relativity not only are all motions naturally described (on equal footing), but they are also *explained* – an accelerated particle is a *curved* (rather *deformed*) worldline, whereas a particle moving with constant velocity is a *straight* worldline (both particles are worldlines in spacetime).

To imply why accelerated motion is experimentally detected (through the resistance an accelerated particle offers to its acceleration) Minkowski specifically pointed out that "Especially

[7]Undoubtedly, de Broglie had realized the profound depth of Minkowski's discovery of the spacetime structure of the world – the very essence of de Broglie's revolutionary idea stems from Minkowski's successful decoding of the message hidden in the failed experiments to detect absolute uniform motion – that *observers in relative motion have different spaces and times* (which led Minkowski to the realization that that is possible in an absolute four-dimensional world): "We were then led to consider the phase wave as due to the intersections by the present space of the fixed observer, of the past, present and future spaces of the moving observer" (this book, p. 41; de Broglie's detailed explanation is on p. 20).

the concept of *acceleration* acquires a sharply prominent character."[8] As Minkowski regarded spacetime and the worldlines (or rather the worldtubes) of particles as representing the real physical world, a real worldtube is expected to resist its deformation; so it does appear that Minkowski's spacetime approach *explains* the sharp distinction between accelerated and uniform (inertial) motion – an accelerated particle resists its acceleration because its worldtube is *deformed* (and therefore resists its deformation), whereas an uniformly moving particle does not resist its motion because its worldline is *not* deformed.

In today's spacetime physics, fully employing Minkowski's approach, the status of acceleration in special relativity is clear – as special relativity applies to flat spacetime, it deals with all motions in the absence of gravitation (= flat spacetime) – inertial and non-inertial (accelerated).

---

[8] H. Minkowski, "Space and Time" in: Hermann Minkowski, *Spacetime: Minkowski's Papers on Spacetime Physics.* Translated by Gregorie Dupuis-Mc Donald, Fritz Lewertoff and Vesselin Petkov. Edited by V. Petkov (Minkowski Institute Press, Montreal 2020), pp. 57-76, p. 66.

## Editor's Note 2

on the following assertion by de Broglie on p. 34:

> It is therefore identical to the momentum and the Maupertuisian action integral has the simple form proposed by Maupertuis himself with the only difference that *mass now varies with velocity* according to Lorentz' law (italics added).

Here de Broglie explicitly regards mass as varying with velocity. Since in recent decades there have been attempts to reject the concept of relativistic (velocity-dependent) mass, it should be emphasized that de Broglie's position is correct, because it reflects the actual situation in spacetime physics – both mass and relativistic mass appear to be equally supported by the experimental evidence – since mass is defined as the measure of the resistance a particle offers to its acceleration (which is the accepted definition based on the experimental evidence) and since it is also an experimental fact that a particle's resistance to its acceleration increases indefinitely (in a given reference frame) as the particle's velocity approaches the speed of light (in the same reference frame), it follows that the particle's mass increases when its velocity increases. Therefore the concept of relativistic mass (like the concept of mass) reflects an experimental fact.

Some authors point out that $\gamma = 1/\sqrt{1-\beta^2}$ should not be "attached" to the mass, because it comes from the 4-velocity. That is, of course, correct – $\gamma$ ensures that the velocity of a particle cannot exceed that of light; in other words, $\gamma$ ensures that no 4-velocity vector, which is timelike, can become lightlike or spacelike. But that is *kinematics*; it says nothing about *dynamics*, that is, it says nothing about why a particle cannot exceed the velocity of light; what is the mechanism that prevents it from doing so. That mechanism is suggested by Newtonian mechanics, where mass is defined as the measure of the *resistance* a particle offers to its acceleration – when Einstein postulated that the velocity of light $c$ is the greatest velocity a particle (with non-zero rest mass) can achieve, it was almost self-evident to assume that a particle would offer an *increasing resistance* when accelerated to velocities approaching that of light, that is, a particle's mass will increase and will approach infinity when the particle's velocity approaches $c$. And that was repeatedly experimentally confirmed. However, increased resistance and increased relativistic mass are rather only naming the mechanism that prevents a particle from reaching the velocity of light; the origin and nature of the resistance a particle offers when accelerated (an open question in classical physics) and of the *increased* resistance a particle offers when accelerated to velocities approaching that of light (an open question in spacetime physics) constitute one of the deepest open questions in spacetime physics.

# 3 The Quantum Conditions of Trajectory Stability

## I. The Bohr-Sommerfeld Stability Conditions

In his theory of the atom, Bohr was the first to suggest that among the closed trajectories that an electron can describe around a positive center, only some are stable, the others being impractical in nature or at least so unstable that they should not be taken into account. Limiting himself to circular trajectories involving only one degree of freedom, Bohr stated the following condition: "Only circular trajectories for which the momentum is an integer multiple of $h/2\pi$ are stable, $h$ being Planck's constant". This can be written as follows:

$$m_0 \omega \mathbf{R}^2 = n\frac{h}{2\pi} \qquad (n \text{ is integer})$$

or:

$$\int_0^{2\pi} p_\theta d\theta = nh$$

$\theta$ being the azimuth chosen as the $q$ Lagrange coordinate, $p_\theta$ the corresponding momentum.

To extend this statement to cases involving several degrees of freedom, both Mr. Sommerfeld and Mr. Wilson have shown that it is generally possible to choose $q_i$i coordinates, such that the conditions for quantifying the orbits are:

$$\oint p_i \, dq_i = n_i h \qquad (n_i \text{ are integers});$$

the symbol $\oint$ indicating an integral extended to the whole range of variation of the coordinate.

In 1917, Mr. Einstein gave the quantization condition an invariant form with respect to coordinate changes.[1] We will state it for the case of *closed* trajectories; it is then as follows:

$$\oint \sum_1^3 p_i \, dq_i = nh \qquad (n \text{ is integer})$$

the integral being extended to the whole length of the trajectory. We recognize the Maupertuisian action integral whose role thus becomes crucial in quantum theory. This integral does not depend on the choice of space coordinates according to a known property that expresses the covariant character of the components $p_i$ of the momentum vector. It is defined by the classical Jacobi method as a complete integral of the partial derivative equation:

$$H\left(\frac{\partial s}{\partial q_i}, q_i\right) = W \qquad i = 1,2\ldots f.$$

complete integral that contains $f$ arbitrary constants, one of which is the energy $W$. If there is only one degree of freedom, Einstein's relation determines the energy $W$; if there is more than one (and in the most important usual case, that of the motion of the electron in the intra-atomic field, there are *a priori* 3), one obtains only a relation between $W$ and the integer $n$; this is what happens for the Keplerian ellipses if one neglects the mass variation with velocity. But if the motion is quasi-periodic, which moreover always occurs because of the above mentioned variation, it is possible to find coordinates that oscillate between limit values

[1] Zum quantensatz von Sommerfeld und Epstein (*Ber. der deutschen. Phys. Ges.*, 1917, p. 82).

(librations) and there is an infinity of pseudo-periods approximately equal to integer multiples of the libration periods. At the end of each of these pseudo-periods, the moving entity has returned to a state as close as one wants to the initial state. Einstein's equation applied to each of these pseudo-periods leads to an infinity of conditions that are compatible only if the multiple Sommerfeld conditions are verified; these being equal in number to the number of degrees of freedom, all the constants are determined and no indeterminacy remains.

For calculation of the Sommerfeld integrals, the Jacobi equation and the residue theorem as well as the concept of the angular variables were successfully used. These questions have been the subject of much work in recent years and are summarized in Sommerfeld's beautiful book "*Atombau und Spectrallinien*" (French edition, Bellenot translation, Blanchard éditeur, 1923). We will not insist on it here and will limit ourselves to noting that, in the end, the problem of quantification comes down entirely in principle to Einstein's condition for closed orbits. If we succeed in interpreting this condition, we will at the same time have shed light on the whole question of stable trajectories.

## II. Interpretation of Einstein's Condition

The notion of phase wave will allow us to provide an explanation of Einstein's condition. It results from the considerations of Chapter II that the trajectory of the moving entity is one of the rays of its phase wave, the latter must run along the trajectory with a constant frequency (since the total energy is constant) and a variable velocity whose value we have learned to calculate. Propagation is therefore analogous to that of a liquid wave in a channel closed on itself and of variable depth. It is physically evident that, to have a stable regime, the length of the channel must be in resonance with the wave; in other words, the wave portions that follow each other at a distance equal to an integer multiple of the length $l$ of the channel and which are therefore at the same point in the channel, must be in phase. The resonance condition is $l = n\lambda$ if the wavelength is constant and $\oint \frac{\nu}{V} dl = n$ (integer) in the general case.

The integral that intervenes here is that of Fermat's principle; however, we have shown that it should be considered equal to the Maupertuisian integral of action divided by $h$. The resonance condition is thus identical to the stability condition required by quantum theory.

This fine result, which is so immediately demonstrable when the ideas of the previous chapter are admitted, is the best justification we can give for our way of tackling the quanta problem.

In the particular case of circular trajectories in the Bohr atom, we obtain

$$m_0 \oint v dl = 2\pi R m_0 v = nh$$

or, since $v = \omega R$ $\omega$ is the angular velocity,

$$m_0 \omega R^2 = n \frac{h}{2\pi}.$$

This is indeed the simple form originally considered by Bohr.

So we can see why some orbits are stable, but we still do not know how the transition from one stable orbit to another takes place. The perturbed regime that accompanies this passage can only be studied with the help of a properly modified electromagnetic theory, and we do not yet have it.

## III. Sommerfeld Conditions for Quasi-Periodic Motions

I propose to demonstrate that, if the stability condition for a *closed orbit* is

$$\oint \sum_1^3 p_i \, dq^i = nh,$$

the stability conditions for quasi-periodic motions are necessarily

$$\oint p_i \, dq^i = n_i h \qquad (n_i \text{ is integer, } i = 1,2,3).$$

The multiple Sommerfeld conditions will thus also be reduced to phase wave resonances.

We must first note that the electron having finite dimensions, if, as we admit, the conditions of stability depend on the reactions exerted on it by its own phase wave, there must be phase agreement between all portions of the wave passing at a distance from the center of the electron less than a given small but *finite* value of the order for example of its radius ($10^{-13}$ cm.). Not to admit this proposition would be tantamount to saying: the electron is a dimensionless geometrical point and the ray of its phase wave is a line of zero thickness. This is physically unacceptable.

Let us now recall a known property of quasi-periodic trajectories. If $M$ is the position of the center of the moving body at a given instant on the trajectory and if one traces from $M$ as the center an arbitrarily chosen, small but finite sphere of radius $R$, it is possible to find an infinite number of time intervals such that at the end of each of them the moving body has returned to the sphere of radius $R$. Moreover, each of these time intervals or "approximate periods" $\tau$ will be able to satisfy the relations:

$$\tau = n_1 T_1 + \epsilon_1 = n_2 T_2 + \epsilon_2 = n_3 T_3 + \epsilon_3,$$

where $T_1$, $T_2$ and $T_3$ are the periods of variation (libration) of the coordinates $q^1 q^2$ and $q^3$. The quantities $\epsilon_i$ can always be made smaller than a certain predefined quantity $\eta$ small but finite. The smaller $\eta$ is chosen, the longer the shorter period $\tau$ will be.

Suppose that the radius $R$ is chosen equal to the maximum distance of action of the phase wave on the electron, distance defined above. Then, we can apply to each approached period $\tau$ the phase agreement condition in the form:

$$\int_o^\tau \sum_1^3 p_i dq^i = nh$$

that can also be written:

$$\begin{aligned} n_1 \int_o^{T_1} p_1 \dot{q}_1 dt + n_2 \int_0^{T_2} p_2 \dot{q}_2 dt \\ + n_3 \int_0^{T_3} p_3 \dot{q}_3 dt + \epsilon_1 (p_1 \dot{q}_1)_\tau \\ + \epsilon_2 (p_2 \dot{q}_2)_\tau + \epsilon_3 (p_3 \dot{q}_3)_\tau = nh. \end{aligned}$$

But a resonance condition is never rigorously satisfied. If the mathematician requires for a resonance to be satisfied that a phase difference be exactly equal to $n \times 2\pi$, the physicist must simply write that it is equal to $n 2\pi \pm a$, $a$ being less than a small but finite quantity $\epsilon$ which measures, if I may say so, the margin within which resonance must be considered to be physically realized.

Quantities $p_i$ and $q_i$ remain finite during the motion and there are six quantities $P_i$ and $\dot{Q}_i$ such that we always have

$$p_i < P_i \qquad \dot{q}_i < \dot{Q}_i \qquad (i = 1, 2, 3).$$

Let's choose the limit $\eta$ such that

$$\eta \sum_1^3 P_i \dot{Q}_i < \frac{\epsilon h}{2\pi};$$

we see that when writing the resonance condition for any of the approximate periods, it will be allowed to disregard the terms in

$\epsilon_i$ and write:

$$n_1 \int_0^{T_1} p_1 \dot{q}_1 dt + n_2 \int_0^{T_2} p_2 \dot{q}_2 dt + n_3 \int_0^{T_3} p_3 \dot{q}_3 dt = nh.$$

In the first member, $n_1, n_2, n_3$, are known integers; in the second member, $n$ is any integer. We have an infinity of similar equations with different values of $n_1$, $n_2$ and $n_3$. To satisfy this condition, it is both required and sufficient that each of the integrals

$$\int_0^{T_i} p_i q_i \, dt = \oint p_i \, dq_i$$

be equal to an integer multiple of $h$.

These are effectively the Sommerfeld conditions.

The previous demonstration seems rigorous. However, there is one objection that needs to be considered. The conditions of stability can only come into play after a period of time of the shortest of the time intervals $\tau$ which are already very long; if we had to wait millions of years, for example, for them to come into play, they would never manifest themselves. This objection is unfounded because the periods $\tau$ are very large compared to the libration periods $T_i$, but can be very small compared to our usual time scale; in the atom, the $T_i$ periods are, in fact, of the order of $10^{-15}$ to $10^{-20}$ seconds.

One can realize the order of magnitude of the approximate periods in the case of the Sommerfeld $L_2$ trajectory for hydrogen. The rotation of the perihelion during a libration period of the vector ray is of the order of $10^{-5} \times 2\pi$. The shortest of the approximated periods would thus be of the order of $10^5$ times the period of the radial variable ($10^{-15}$ seconds), i.e. of the order of $10^{-10}$ seconds. It thus seems that the stability conditions will come into play in a time inaccessible to our experience and, consequently, that the "non-resonant" trajectories will seem to us to be non-existent.

The principle of the demonstration developed above was borrowed from Mr. Léon Brillouin who wrote in his thesis (p. 351):

For the Maupertuis integral taken over all the approximate periods $\tau$ to be an integer multiple of $h$, each of the integrals relative to each variable and taken over the corresponding period must be equal to an integer number of quanta; this is effectively how Sommerfeld writes his quanta conditions.

# 4 Quantification of the Simultaneous Motions of two Electrical Centers

## I. Difficulties Raised by this Issue

In the previous chapters, we have constantly considered an "isolated piece" of energy. This expression is clear when it relates to an electric corpuscle (proton or electron) far from any other electrified body. But if electrified centers are interacting, the concept of an isolated piece of energy becomes less clear. There is a difficulty here that is by no means peculiar to the theory contained in the present work and which is not elucidated in the present state of the dynamics of the Relativity theory.

To fully understand this difficulty, let us consider a proton (hydrogen nucleus) of rest mass $M_0$ and an electron of rest mass $m_0$. If these two entities are very distant from each other so that their interaction is negligible, the principle of the inertia of energy applies without difficulty: the proton has the internal energy $M_0c^2$ and the electron $m_0c^2$. The total energy is therefore $(M_0 + m_0)c^2$. But if their two centers are close enough to each other to take into account their mutual potential energy $-P(< 0)$ then how will the idea 0f the inertia of energy be expressed? The total energy being obviously $(M_0 + m_0)c^2 - P$, can we admit that the proton always has a rest mass $M_0$ and the electron a rest mass $m_0$? On the contrary, should we share the potential energy between the two constituents of the system, attributing to the electron a rest mass $m_0 - \alpha P / c^2$ and to the proton a rest mass $M_0 - (1 - \alpha) P / c^2$ ? In this case, what is the value of $\alpha$ and how does this quantity depend on $M_0$ and $m_0$ ?

In the theories of Bohr and Sommerfeld on the atom, it is admitted that the electron always has its rest mass $m_0$ whatever its position in the electrostatic field of the nucleus. Since the potential energy is always much smaller than the internal energy $m_0 c^2$, this hypothesis is about right, but nothing says that it is rigorous. One can easily calculate the order of magnitude of the maximum correction (corresponding to $\alpha = 1$) that would have to be made to the value of the Rydberg constant for the different terms of the Balmer series if one adopted the opposite hypothesis. We find $\frac{\delta R}{R} = 10^{-5}$. This correction would therefore be much smaller than the difference between the Rydberg constants for hydrogen and for helium $\left(\frac{1}{2000}\right)$, a difference that Bohr remarkably accounted for by considering the drag of the nucleus. However, given the extreme precision of spectroscopic measurements, it is perhaps permissible to think that the variation of the Rydberg constant due to the variation of the electron rest mass as a function of its potential energy could be highlighted if it exists.

## II. The Drag of the Nucleus in the Hydrogen Atom

Closely related to the previous one is the question of how to apply quantum conditions to a set of relatively moving electrical centers. The simplest case is that of the motion of the electron in the hydrogen atom when the simultaneous displacements of the nucleus are taken into account. Dr. Bohr was able to deal with this problem using the following theorem from Classical Mechanics: "If we relate the motion of the electron to axes of fixed directions related to the nucleus, this motion is the same as if these axes were Galilean and if the electron had a mass $\mu_0 = \frac{m_0 M_0}{m_0 + M_0}$."

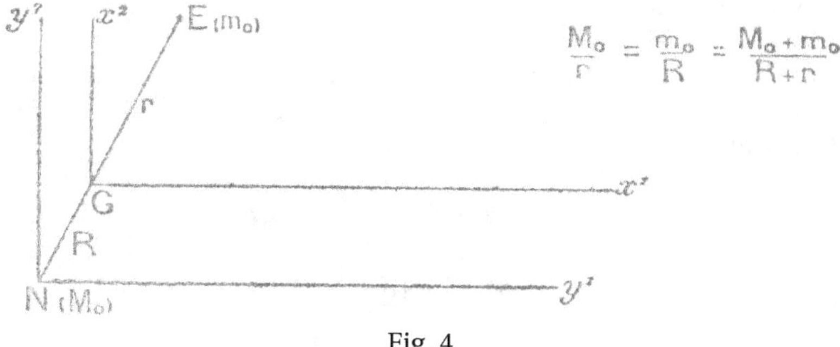


Fig. 4

In the system of axes linked to the nucleus, the electrostatic field acting on the electron can be considered as constant at any point in space and we are thus brought back to the problem without motion of the nucleus thanks to the substitution of the fictitious mass $\mu_0$ to the real mass $m_0$. In chapter II of this work, we have established a general parallelism between the fundamental quantities of Dynamics and those of Wave Theory; the theorem stated above thus determines which values should be attributed to the frequency of the electronic phase wave and to its velocity in the system linked to the nucleus, a system that is not Galilean. Thanks to this technique, the quantum conditions of stability can also be considered in this case as being interpretable via the resonance of the phase wave. We will clarify by focusing on the case in which the nucleus and the electron describe circular orbits around their common center of gravity. The plane of these orbits will be taken as the plane of the coordinates of indices 1 and 2 in both systems. The space coordinates in the Galilean system related to the center of gravity will be $x^1, x^2$ and $x^3$, those of the system related to the nucleus will be $y^1, y^2$ and $y^3$. Finally we will have $x^4 = y^4 = ct$. Let us name $\omega$ the rotation velocity of line $NE$ about point $G$. Let us pose by definition:

$$\eta = \frac{M_0}{m_0 + M_0}.$$

The formulas that allow switching from one axis system to the other are as follows:

$$y^1 = x^1 + R\cos\omega t \qquad y^2 = x^2 + R\sin\omega t \qquad y^3 = x^3 \qquad y^4 = x^4.$$

Consequently:

$$\begin{aligned} ds^2 &= (dx^4)^2 - (dx^1)^2 - (dx^2)^2 - (dx^3)^2 \\ &= \left(1 - \frac{\omega^2 R^2}{c^2}\right)(dy^4)^2 - (dy^1)^2 - (dy^2)^2 \\ &- 2\frac{\omega R}{c}\sin\omega t\, dy^1\, dy^4 + 2\frac{\omega R}{c}\cos\omega t\, dy^2\, dy^4. \end{aligned}$$

The components of the four-momentum vector are defined by the following relations:

$$u^i = \frac{dy^i}{ds} \qquad p_i = m_0\, c\, u_i + e\phi i = m_0\, g_{ij}\, u^j + e\phi_i.$$

We easily find:

$$\begin{aligned} p_1 &= \frac{m_0}{\sqrt{1-\eta^2\beta^2}}\left[\frac{dy_1}{dt} + \omega R\sin\omega t\right] \\ p_2 &= \frac{m_0}{\sqrt{1-\eta^2\beta^2}}\left[\frac{dy^2}{dt} - \omega R\cos\omega t\right] \\ p_3 &= 0. \end{aligned}$$

The resonance of the phase wave is expressed according to the general ideas of Chapter II by the relation:

$$\left[\oint \frac{1}{h}(p_1 dy^1 + p_2 dy^2)\right] = n \qquad (n \text{ is integer})$$

the integral being extended to the circular trajectory of radius $R + r$ travelled by the electron about the nucleus.

Since we have:

$$\frac{dy^1}{dt} = -\omega(R+r)\sin\omega t \qquad \frac{dy^2}{dt} = \omega(R+r)\cos\omega t$$

we obtain:

$$\frac{1}{h}\oint (p_1 dy^1 + p_2 dy^2) = \frac{1}{h}\oint \frac{m_0}{\sqrt{1-\eta^2\beta^2}}\,(v dl - \omega R\, v\, dt)$$

by naming $v$ the velocity of the electron with respect to the $y$-axis and by $dl$ the length element of its trajectory,

$$v = \omega(R+r) = \frac{dl}{dt}.$$

Finally, the resonance condition becomes:

$$\frac{m_0}{\sqrt{1-\eta^2\beta^2}}\,\omega(R+r)\left(1-\frac{\omega R}{v}\right)2\pi(R+r) = nh$$

or, assuming with classical mechanics that $\beta^2$ is negligible in front of the unit,

$$2\pi m_0 \frac{M_0}{m_0+M_0}\,\omega(R+r)^2 = nh.$$

This is indeed Bohr's formula which can be deduced from the theorem stated above and which can therefore be considered here again as a resonance condition of the electronic wave belonging to the system linked to the nucleus of the atom.

## III. The two Phase Waves of the Nucleus and the Electron

In what preceded, the introduction of axes linked to the nucleus somehow allowed eliminating its motion in the system and to consider the motion of the electron in a constant electrostatic field; we were thus brought back to the problem discussed in Chapter II. But, if we move on to other axes related, for example, to the center of gravity, the nucleus and the electron will both describe closed trajectories, and the ideas that guided us so far must inevitably lead us to conceive of the existence of two phase waves: that of the electron and that of the nucleus; we must examine how the resonance conditions of these two waves should be expressed and why they are compatible. Let's first consider the phase wave of the electron. In the system linked to the nucleus, the resonance condition for this wave is:

$$\oint p_1 dy^1 + p_2 dy^2 = 2\pi \frac{m_0 M_0}{m_o + M_0} \omega (R+r)^2 = nh,$$

the integral being taken at *constant time* along the circle of center $N$ and radius $R+r$, which are the relative trajectory of the mobile and the radius of its wave. If we switch to the axes related to point $G$, the relative trajectory becomes a circle of center $G$ and radius $r$; the radius of the phase wave passing through $E$ is at *each instant* the circle of center $N$ and radius $R+r$, but this circle is mobile because its center translates with a uniform motion about the origin of the coordinates. The resonance condition of the electronic wave at a given instant is not modified; it is still written:

$$2\pi \frac{m_0 M_0}{m_0 + M_0} \omega (R+r)^2 = nh.$$

Let's move on to the nucleus wave. In all the above, the nucleus and the electron play a perfectly symmetrical role and the resonance condition must be obtained by inverting $M_0$ and $m_0$, $R$ and $r$. So we return to the same formula.

In summary, we can see that Bohr's condition can be interpreted as the expression of the resonance of each of the participating waves. The stability conditions for the motions of the nucleus and the electron considered in isolation are compatible because they are identical.

It is instructive to plot in the axes system linked to the center of gravity the rays at instant $t$ of the two phase waves (solid line) and the trajectories described over time by the two moving entities (dotted line). We can then imagine how each moving entity describes its trajectory with a velocity that at any moment is tangent to the phase wave radius.

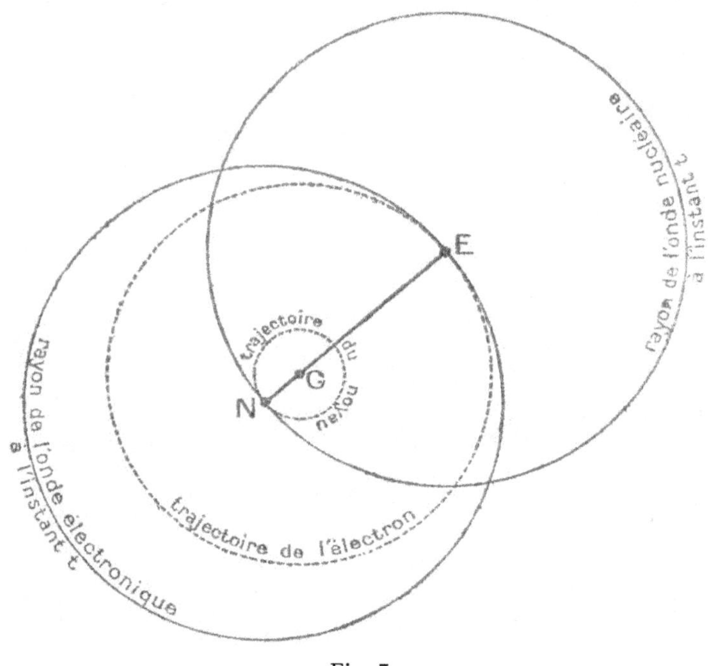


Fig. 5

Let us insist on one last point. The rays of the wave at instant $t$ are the envelopes of the velocity of propagation, but these rays are not the trajectories of the energy, they are only tangent to them at each point. This reminds us of the known conclusions of hydrodynamics where the lines of flow, envelopes of velocities, are the trajectories of fluid particles only if their shape is invariant, in other words, if the motion is permanent.

# 5 THE QUANTA OF LIGHT[1]

## I. The Atom of Light

As mentioned in the introduction, the development of radiation physics has been going on for several years in the direction of at least a partial return to the corpuscular theory of light. An attempt made by us to develop an atomic theory of black radiation, published by the *Journal de Physique* in November 1922 under the title "Quanta de lumière et rayonnement noir", the main results of which will be given in Chapter VII, reinforced our idea of the real existence of the light atom. The ideas presented in Chapter I, and whose deduction of the conditions of stability in the Bohr atom in Chapter III seems to provide such an interesting confirmation, seem to lead us into taking a small step towards the synthesis of Newton's and Fresnel's conceptions.

Without concealing the difficulties raised by such boldness, we will try to clarify how the light atom can currently be represented. We conceive it in the following way: for an observer connected to it, it appears as a small region of space around which energy is very strongly condensed and forms an indivisible whole. This agglomeration of energy having for total value $\epsilon_0$ (measured by the bound observer), it is necessary, according

[1]See A. Einstein, "Über einem die Erzeugung und Verwandlung des Lichtes betreffenden heuristischen Gesichtspunkt," *Ann. der Phys.*, **17**, 132 (1905); *Phys. Zeitsch.*, **10**, 185 (1909).

to the principle of the inertia of energy, to attribute to it a proper mass:

$$m_0 = \frac{\epsilon_0}{c^2}.$$

This definition is entirely equivalent to the definition of the electron. There remains however an essential difference of structure between the electron and the light atom. While the electron must be up to now considered as endowed with a spherical symmetry, the light atom must have an axis of symmetry corresponding to polarization. We will thus imagine the quantum of light as having the same symmetry as a doublet in the electromagnetic theory. This representation is quite provisional and we will only be able to specify with any chance of accuracy the constitution of the luminous unit after having made the electromagnetism undergo deep modifications and this work is not accomplished.

In accordance with our general ideas, we will suppose that in the very constitution of the quantum of light there exists a periodic phenomenon whose natural frequency $\nu_0$ is given by the relation:

$$\nu_0 = \frac{1}{h} m_0 c^2.$$

The phase wave corresponding to the motion of this quantum with velocity $\beta c$ will have as frequency:

$$\nu = \frac{1}{h} \frac{m_0 c^2}{\sqrt{1-\beta^2}}$$

and it is quite appropriate to assume that this wave is identical to that of the wave theories, or more exactly that the classically conceived distribution of waves in space is a kind of average in time of the real distribution of phase waves accompanying light atoms.

It is an experimental fact that light energy travels with a velocity indistinguishable from the limit value $c$. Since the velocity $c$ is a speed that energy can never reach because of the law of mass variation with velocity, we are naturally led to suppose that radiation is formed by light atoms moving with velocities very close to $c$, but slightly lower.[1]

If a body has an extraordinarily small proper mass, in order to communicate an appreciable kinetic energy to it, it will have to be given a velocity very close to $c$; this results from the expression of kinetic energy:

$$E = m_0 c^2 \left( \frac{1}{\sqrt{1-\beta^2}} - 1 \right).$$

Moreover, all energies, i.e., from 0 to $\infty$, correspond to velocity values within a very small interval $c - \epsilon, c$. We therefore conceive that, assuming $m_0$ to be extraordinarily small (we will specify later), light atoms with appreciable energy will all have a velocity very close to $c$ and, despite the almost equality of their velocities, will have very different energies.

[1]EDITOR'S NOTE: It is now accepted that light atoms (which we now call photons) have zero rest mass and travel exactly with the speed $c$. Or, stated rigorously, photons are null geodesics in spacetime since nothing travels is spacetime, where particles and photons are a forever given web of worldlines (containing their entire histories in time *at once*, i.e., *en bloc*, like in a film strip); in Minkowski's best explanation: "The whole world presents itself as resolved into such worldlines, and I want to say in advance, that in my understanding the laws of physics can find their most complete expression as interrelations between these worldlines" [1] (see also [2]). Some physicists sometimes say it is a matter of description whether we use four-dimensional or three-dimensional presentations of the world. However, whether the physical world is four-dimensional or three-dimensional is not a matter of description.

1. H. Minkowski, "Space and Time" in: Hermann Minkowski, *Spacetime: Minkowski's Papers on Spacetime Physics.* Translated by Gregorie Dupuis-Mc Donald, Fritz Lewertoff and Vesselin Petkov. Edited by V. Petkov (Minkowski Institute Press, Montreal 2020), pp. 57-76, pp. 58-59.

2. Geroch also expressed it explicitly and exactly: "There is no dynamics in space-time: nothing ever happens there. Space-time is an unchanging, once-and-for-all picture encompassing past, present, and future," Robert Geroch, *General Relativity: 1972 Lecture Notes* (Minkowski Institute Press, Montreal 2013), pp. 4-5.

Since we equate the phase wave with the classical light wave, the frequency $\nu$ of the radiation will be defined by the relation:

$$\nu = \frac{1}{h}\frac{m_0 c^2}{\sqrt{1-\beta^2}}.$$

Let's note, a fact that must be remembered every time we are dealing with atoms of light, the extreme smallness of $m_0 c^2$ in front of $\frac{m_0 c^2}{\sqrt{1-\beta^2}}$; kinetic energy can therefore be written here simply as follows:

$$\frac{m_0 c^2}{\sqrt{1-\beta^2}}.$$

The light wave of frequency $\nu$ would thus correspond to the displacement of a light atom with the velocity $v = \beta c$ connected to $\nu$ by the relation:

$$v = \beta c = c\sqrt{1 - \frac{m_0^2 c^4}{h^2 \nu^2}}.$$

Except for extremely slow vibrations, $\frac{m_0 c^2}{h\nu}$, and all the more so its square, will be very small and we can pose:

$$v = c\left(1 - \frac{m_0^2 c^4}{2h^2 \nu^2}\right).$$

We can try to set an upper limit for the value of $m_0$. Indeed, experiments by means of wireless telegraphy have shown that radiations of a few kilometers wavelength still substantially propagate with velocity $c$. Let us suppose that waves for which $\frac{1}{\nu} = 10^{-4}$ seconds have a velocity different from $c$ by less than one hundredth. The upper limit of $m_0$ will be:

$$(m_0)_{max} = \frac{\sqrt{2}}{10}\frac{h\nu}{c^2}$$

or approximately $10^{-44}$ grams. It is even likely that $m_0$ should be chosen even smaller; perhaps, one can hope that one day, by measuring the vacuum velocity of very low frequency waves, one will find numbers quite significantly lower than $c$.

It must be remembered that the propagation velocity just mentioned is not that of the phase wave, always higher than $c$, but that of the energy displacement, the only one experimentally detectable.[2]

## II. The Motion of the Atom of Light

The atoms of light for which $\beta = 1$ substantially, would thus be accompanied by phase waves whose velocity $\frac{c}{\beta}$ would also be substantially equal to $c$; this is, we think, the coincidence that would establish between the atom of light and its phase wave a particularly narrow place that would constitute the double corpuscular and undulatory aspect of the radiation. The identity of the Fermât and the least action principles would explain why the rectilinear propagation of light is compatible with both points of view.

The trajectory of the light corpuscle would be one of the rays of its phase wave. There is reason to believe, as we will see later, that several corpuscles could have the same phase wave; their trajectories would then be different rays of this wave. The old idea that the ray is the trajectory of the energy would thus be confirmed and clarified.

However, rectilinear propagation is not an absolutely general fact; a light wave falling on the edge of a screen diffracts and penetrates the geometric shadow, the rays that pass at small distances from the screen in relation to the wavelength are deflected and no longer follow Fermât's law. From the undulatory point of view, the deflection of the rays is explained by the imbalance introduced between the actions of the various areas very

[2]Regarding the objections raised by the ideas contained in this paragraph, see the Appendix.

close to the wave due to the presence of the screen. From the opposite point of view, Newton assumed a force exerted by the edge of the screen on the corpuscle. It seems that we can arrive at a synthesized view: the ray of the wave would bend as predicted by the theory of oscillations and the moving entity for which the principle of inertia would no longer be valid, would undergo the same deflection as the ray to which its motion is related; perhaps we could say that the wall exerts a force on the corpuscle, a force applied to it if we take the curvature of the trajectory as the criterion for the existence of a force.

In the above we have been guided by the idea that the corpuscle and its phase wave are not different physical realities. If we think about it we will see that it seems to lead to the following conclusion: "Our dynamics (including its Einsteinian form) has lagged behind Optics: it is still at the stage of Geometric Optics". If it seems to us today quite probable that all waves have energy concentrations, on the other hand the dynamics of the material point probably conceals a wave propagation and the true meaning of the principle of least action is to express a phase concordance.

It would be very interesting to look for the interpretation of diffraction in space-time, but here we meet the difficulties pointed out in Chapter II about non-uniform motion and we could not clarify this issue in a satisfactory way.

## III. Some Agreements between the Opposing Radiation Theories

We will show with a few examples how easily the corpuscular radiation theory accounts for a number of known results of wave theories.

a) Doppler effect by motion of the source:

Let's consider a moving light source with velocity $v = \beta c$ in the direction of an observer who is assumed to be stationary. This source is assumed to emit light atoms, the phase wave frequency is $\nu$ and its velocity $c(1-\epsilon)$ where

$$\epsilon = \frac{1}{2}\frac{m_0^2 c^4}{h^2 \nu^2}.$$

For the stationary observer, these quantities have for values $\nu'$ and $c(1-\epsilon')$. The velocities addition theorem gives:

$$c(1-\epsilon') = \frac{c(1-\epsilon)+v}{1+\dfrac{c(1-\epsilon)\,v}{c^2}},$$

where

$$1-\epsilon' = \frac{1-\epsilon+\beta}{1+(1-\epsilon)\beta}$$

or, by neglecting $\epsilon\epsilon'$:

$$\frac{\epsilon}{\epsilon'} = \frac{\nu'^2}{\nu^2} = \frac{1+\beta}{1-\beta}, \qquad \frac{\nu'}{\nu} = \sqrt{\frac{1+\beta}{1-\beta}}$$

if $\beta$ is small, we find the formulas of the old optics:

$$\frac{\nu'}{\nu} = 1+\beta, \qquad \frac{T'}{T} = 1-\beta = 1-\frac{v}{c}.$$

It is also easy to find the ratio of emitted intensities for the two observers. During the unit of time, the moving observer sees the source emitting $n$ atoms of light per unit area. The energy density of the beam evaluated by this observer is thus $nh\upsilon/c$ and the intensity is $I = nh\upsilon$. For the stationary observer, the $n$ atoms are emitted in a time equal to $\frac{1}{\sqrt{1-\beta^2}}$ and they fill a volume

$$c(1-\beta)\frac{1}{\sqrt{1-\beta^2}} = c\sqrt{\frac{1-\beta}{1+\beta}}.$$

The energy density of the beam seems to him to be:

$$\frac{nh\nu'}{c}\sqrt{\frac{1+\beta}{1-\beta}}$$

and its intensity:

$$I' = nh\nu'\sqrt{\frac{1+\beta}{1-\beta}} = nh\nu'\,\frac{\nu'}{\nu}.$$

whence:

$$\frac{I'}{I} = \left(\frac{\nu'}{\nu}\right)^2.$$

All these formulas are demonstrated from an undulatory point of view in Laue's book, *Die Relativitätstheorie*, volume I, 3rd ed, p. 119.

b) Reflection on a moving mirror:

Let's consider the reflection of light corpuscles normally falling on a perfectly reflecting plane mirror that moves with velocity $\beta c$ in a direction perpendicular to its surface.

For the stationary observer, $\nu_1'$ is the frequency of the phase waves accompanying the incident corpuscles and $c(1-\epsilon_1')$ their velocity. The same quantities for the linked observer will be $\nu_1$ and $c(1-\epsilon_1)$.

If we consider reflected corpuscles, the corresponding values will be referred to as$\nu_2$, $c(1-\epsilon_2)$, $\nu_2'$, and $c(1-\epsilon_2')$. The composition of the velocities gives:

$$c(1-\epsilon_i) = \frac{c(1-\epsilon_1')+\beta c}{1+\beta(1-\epsilon_1')}, \qquad c(1-\epsilon_2) = \frac{c(1-\epsilon_2')-\beta c}{1-\beta(1-\epsilon_2')}.$$

For the linked observer, there is reflection on a fixed mirror without any change of frequency since the energy is conserved. Hence:

$$\nu_1 = \nu_2, \qquad \epsilon_1 = \epsilon_2, \qquad \frac{1-\epsilon_1'+\beta}{1+\beta(1-\epsilon_1')} = \frac{1-\epsilon_2'-\beta}{1-\beta(1-\epsilon_2')}.$$

Neglecting $\epsilon'_1, \epsilon'_2$ we obtain:

$$\frac{\epsilon'_1}{\epsilon'_2} = \left(\frac{\nu'_2}{\nu'_1}\right)^2 = \left(\frac{1+\beta}{1-\beta}\right)^2 \quad \text{or} \quad \frac{\nu'_2}{\nu'_1} = \frac{1+\beta}{1-\beta}.$$

If $\beta$ is small, we recover the classical formula:

$$\frac{T_2}{T_1} = 1 - 2\frac{v}{c}.$$

It would be easy to deal with the problem by assuming an oblique incidence.

Let us designate by $n$ the number of corpuscles reflected by the mirror during a given time. The total energy of the $n$ corpuscles after reflection $E'_2$ is at their total energy before reflection $E'_1$ in the ratio:

$$\frac{nh\nu'_2}{nh\nu'_1} = \frac{\nu'_2}{\nu'_1}.$$

Electromagnetic theory also gives this relationship, but here it is quite obvious. If the $n$ corpuscles occupied before reflection a volume $V_1$, they will occupy after reflection a volume

$$V_2 = V_1 \frac{1-\beta}{1+\beta}$$

as shown by a very simple geometrical reasoning. The intensities $I'_1$ and $I'_2$ before and after reflection thus have the ratio:

$$\frac{I'_2}{I'_1} = \frac{nh\nu'_2}{nh\nu'_1}\frac{1+\beta}{1-\beta} = \left(\frac{\nu'_2}{\nu'_1}\right)^2.$$

All these results are demonstrated from a wave perspective by Laue, page 124.

c) Radiation pressure of black body radiation:

Let us consider an enclosure filled with black body radiation at temperature $T$. What is the pressure applied to the walls of the enclosure? For us the black body radiation will be a gas of atoms of light and we will assume that the velocity distribution is isotropic. Let $u$ be the total energy (or, which here is the same, the total kinetic energy) of the atoms of light contained in the unit of volume. Let $ds$ be a surface element of the wall, $dv$ a volume element, $r$ their distance, $\theta$ the angle of the line joining them with the normal to the surface element.

The solid angle under which the element $ds$ is seen from point $O$, center of $dv$, is :

$$d\Omega = \frac{ds\cos\theta}{r^2}.$$

Let us consider only those atoms of light of volume $dv$ whose energy lies between $\omega$ and $\omega + d\omega$, in number $n_\omega d\omega dv$; the number of those whose velocity is directed towards $ds$ is due to isotropy:

$$\frac{d\Omega}{4\pi} \times n_\omega d\omega dv = n_\omega d\omega \frac{ds\cos\theta}{4\pi r^2} dv.$$

Taking a spherical coordinate system with the normal to $ds$ as the polar axis, we find:

$$dv = r^2 \sin\theta d\theta d\psi dr.$$

Moreover, the kinetic energy of an atom of light being $\frac{m_0 c^2}{\sqrt{I-\beta^2}}$ and its momentum $G = \frac{m_0 v}{\sqrt{I-\beta^2}}$ with $v = c$ very approximately, we have:

$$\frac{W}{c} = G.$$

Thus, the reflection from the angle $\theta$ of an atom of energy $W$ communicates to $ds$ an impulse

$$2G\cos\theta = 2\frac{W}{c}\cos\theta.$$

The atoms of energy of volume $dv$ having this energy will thus communicate to it by reflection an impulse equal to:

$$2\frac{W}{c}\cos\theta\, n_\omega\, d\omega\, r^2 \sin\theta\, d\theta\, d\psi\, dr \frac{ds\cos\theta}{4\pi r^2}.$$

Let us integrate with respect to $W$ from 0 to $\infty$ noting that

$$\int_0^{+\infty} \omega\, n_\omega\, d\omega = u,$$

with respect to the angles $\psi$ and $\theta$ respectively from 0 to $2\pi$ and from 0 to $\pi/2$, finally with respect to $r$ from 0 to $c$. We obtain the total impulse suffered in one second by the element $ds$ and, dividing by $ds$, the radiation pressure:

$$p = u \int_0^{\frac{\pi}{2}} \cos^2\theta \sin\theta\, d\theta = \frac{u}{3}.$$

The radiation pressure is equal to one third of the energy contained in the unit of volume, a result known from classical theories. The ease with which we have just found in this paragraph certain results also provided by the undulatory conceptions of radiation reveals the existence between the two apparently opposite viewpoints of a secret harmony whose notion of phase wave enables us to glimpse the nature.

## IV. Wave Optics and Light Quanta[3]

The stumbling block of the light quanta theory is the explanation of the phenomena that make up wave optics. The main reason for this is that this explanation requires the intervention of the phase of periodic phenomena, so it may seem that we have taken a very big step forward in understanding a close link between the motion of a corpuscle of light and the propagation of a certain wave. Indeed, it is very likely that, if the theory of light quanta ever succeeds in explaining the phenomena of wave optics, it is by such conceptions that it will do so. Unfortunately, it is still impossible to reach satisfactory results in this respect and

[3] See on this subject Bateman (H.). On the theory of light quanta, *Phil. Mag.*, **46** (1923), 977 where one can find a history and a bibliography.

only the future will tell if Einstein's audacious conception, judiciously adapted and completed, will be able to accommodate in its framework the numerous phenomena whose marvelous precision led 19th century physicists to consider the wave hypothesis as definitively established.

Let us limit ourselves to circling around this difficult problem without trying to attack it head-on. In order to progress along the path followed so far, we would need to establish, as previously mentioned, a certain link of probably statistical nature between the wave conceived in the classical way and the superposition of phase waves; this would certainly lead to attribute an electromagnetic nature to the phase wave and consequently also to the periodic phenomenon defined in chapter 1.

The emission and absorption of radiation can be considered to be proven with near certainty as being discontinuous. Electromagnetism, or more precisely the electron theory, therefore gives us an inaccurate view of the mechanics of these phenomena. However, Dr. Bohr, through his principle of correspondence, taught us that if we consider the predictions of this theory for the radiation emitted by a set of electrons, they undoubtedly possess a sort of overall accuracy. Perhaps the whole electromagnetic theory would have only a statistical value; Maxwell's laws would then appear as a continuous approximation of a discontinuous reality, in much the same way (but only slightly) that the laws of hydrodynamics give a continuous approximation of the very complex and rapidly changing motions of fluid molecules. This idea of correspondence, that still seems rather imprecise and elastic, will have to serve as a guide for bold researchers who will want to constitute a new electromagnetic theory more in line with quantum phenomena than the current theory.

We will reproduce in the following paragraph some of the considerations that we made about interferences; frankly speaking, they should be considered as vague suggestions rather than true explanations.

## V. Interferences and Coherence

We will first ask ourselves how one notices the presence of light at a point in space. We can place a body on which the radiation can have a photoelectric, chemical, calorific, etc. effect, and it is possible that on final analysis all such effects are photoelectric. It is also possible to observe the diffusion of waves produced by matter at the point considered in space. We can therefore say that where radiation cannot react on matter, it is undetectable experimentally. The electromagnetic theory admits that photographic actions (Wiener's experiments) and scattering are related to the intensity of the resulting electric field; where the electric field is zero, if there is magnetic energy, it is undetectable.

The ideas developed here lead to assimilate phase waves to electromagnetic waves, at least as far as the distribution of phases in space is concerned, the issue of intensities remaining to be determined. This idea, together with the idea of correspondence, leads us to think that the probability of reactions between atoms of matter and atoms of light is at each point linked to the resultant (or rather to the average value of the latter) of one of the vectors characterizing the phase wave; there is interference where this resultant is null, light is undetectable. It is therefore conceivable that an atom of light passing through a region where phase waves interfere can be absorbed by matter at certain points and not at others. This is the still very qualitative principle of an explanation of interference compatible with the discontinuity of radiant energy. Mr. Norman Campbell in his book *Modern electrical theory* (1913) seems to have glimpsed a similar solution when he wrote: "Only corpuscular theory can explain how the energy of radiation is transferred from one place to another while only wave theory can explain why the transfer along one path depends on the transfer to another. It almost seems that the energy itself is transported by corpuscles while the power to absorb it and make it perceptible to experience is transported by spherical waves".

For interferences to occur regularly, it seems essential to establish a sort of dependency between the emissions of the various atoms of light from the same source. We proposed to express this dependence by the following postulate. "The phase wave related to the motion of an atom of light can, by passing over excited material atoms, trigger the emission of other atoms of light whose phase will be in agreement with that of the wave". A wave could thus carry many small centers of energy condensation that would slide slightly on its surface while remaining in phase with it. If the number of transported atoms of light were extremely large, the structure of the wave would come close to classical designs as a sort of limit.

## VI. The Bohr Frequency Law. Conclusions

From any viewpoint considered, the detail of the internal transformations undergone by the material atom when it absorbs or when it emits, cannot yet be imagined. Let us still admit the granular hypothesis: we do not know if the quantum absorbed by the material atom somehow merges with it or if it remains inside it as an isolated unit, nor do we know if the emission is the expulsion of a preexisting quantum in the material atom or the creation of a new unit at the expense of its internal energy. In any case, it seems certain that the emission concerns only a single quantum; therefore, the total energy of the corpuscle equal to $h$ times the frequency of the accompanying phase wave should, in order to preserve energy conservation, be equal to the decrease of the total energy content of the material atom and this gives us the Bohr's law of frequencies:

$$h_{\nu} = W_1 - W_2$$

We can therefore see that our conceptions, after having led us to a simple explanation of the conditions of stability, also allow us to obtain the law of frequencies *on the condition, however, that we admit that the emission always relates to a single corpuscle.*

Let us note that the representation of the emission provided by the theory of light quanta seems to be confirmed by the conclusions of Messrs. Einstein and Léon Brillouin[4] who showed the need to introduce into the analysis of the reactions between the black body radiation and a free particle the idea of a strictly directed emission.

What should we conclude from this whole chapter? Certainly this phenomenon of dispersion, which seemed incompatible with the notion of light quanta in its simplistic form, now seems less impossible to reconcile with it thanks to the introduction of a phase. The recent theory of $X$- and $\gamma$-rays scattering, given by M. A.-H. Compton, which we will discuss later, seems to be based on serious experimental evidence and to make tangible the existence of light corpuscles in a field where wave patterns reigned supreme. Nevertheless, it is undeniable that the concept of luminous energy grains still fails to solve the problems of wave optics and that it faces very serious difficulties; it would be premature to say whether or not it will succeed in overcoming them.

---

[4] A. Einstein, *Phys. Zeitschr.*, 18, 121, 1917; L. Brillouin, *Journ. d. Phys.*, série VI, 2, 142, 1921.

# 6 THE SCATTERING OF $X-$ AND $\gamma$-RAYS

## I. The Theory of J. J. Thomson[1]

In this chapter, we want to study the scattering of $X$- and $\gamma$-rays and show with this particularly suggestive example the respective current positions of the electromagnetic theory and of the light quanta theory:

Let's start by defining the very phenomenon of scattering: when a beam of rays is sent onto a piece of matter, part of the energy is generally scattered in all directions. It is said that there is scattering and weakening by diffusion of the beam during the transit through the substance.

The electronic theory interprets this phenomenon very simply. It supposes (which seems to be in direct opposition to Bohr's atomic model) that electrons contained in an atom are subjected to quasi-elastic forces and have a well defined vibratory period. Therefore, the passage of an electromagnetic wave over these electrons will give them an oscillatory motion whose amplitude will generally depend on both the frequency of the incident wave and the natural frequency of the electronic resonators. According to the theory of wave acceleration (la théorie de l'onde

[1] *Passage de l'électriciteé à travers les gaz.* French translation Fric and Faure. Gauthier-Villars, 1912, p. 321.

d'accélération), the motion of the electron will be continuously damped by the emission of a cylindrically symmetrical wave. An equilibrium regime will be established in which the resonator will draw from the incident radiation the necessary energy to compensate for this damping. The final result will be a scattering of a fraction of the incident energy in all directions of space.

To calculate the magnitude of the scattering phenomenon, we must first determine the motion of the vibrating electron. To this effect we must express the equilibrium between the resultant of the inertial force and the quasi-elastic force on the one hand and the electrical force exerted by the incident radiation on the electron on the other hand. In the visible range, examination of the numerical values shows that one can neglect the inertia term in front of the quasi-elastic term and one is thus led to attribute to the amplitude of the vibratory motion a value proportional to the amplitude of the exciting light, but independent of its frequency. The dipole radiation theory then teaches us that the global secondary radiation is inversely related to the 4th power of the wavelength; the higher the frequency, the more diffuse the radiation is. It is on this conclusion that Lord Rayleigh grounded his beautiful theory of the blue color of the sky.[2]

In the domain of very high frequencies ($X$- and $\gamma$-rays), it is on the contrary the quasi-elastic term which is negligible compared to the term inertia. Everything happens as if the electron was free and the amplitude of its vibratory motion was proportional not only to the incident amplitude, but also to the 2nd power of the wavelength. As a result, the overall scattered intensity is this time independent of the wavelength. It was Mr. J. J. Thomson who first drew attention to this fact and constituted the first theory of $X$-ray scattering. The two main conclusions were as follows:

1° If one designates by $\theta$ the angle of the extension of the direction of incidence with the direction of scattering, the scat-

[2]Lord Rayleigh deduced this theory from the elastic concept of light, which is on this point entirely in agreement with the electromagnetic concept.

tered energy varies according to $\theta$ as $\frac{1+\cos^2\theta}{2}$.

2° The total energy scattered by an electron in one second corresponds in relation to the incident intensity to the ratio:

$$\frac{I_\alpha}{I} = \frac{8\pi}{3}\frac{e^4}{m_0^2 c^4}$$

$e$ and $m_0$ being the electron constants, $c$ the speed of light.

An atom certainly contains several electrons; today, there is good reason to believe that their number $p$ is equal to the atomic number of the element. Thomson assumed that the waves emitted by the $p$ electrons in an atom are "inconsistent" and therefore considered the energy emitted by an atom to be $p$ times the energy that would be emitted by a single electron. From an experimental point of view, the scattering results in a gradual weakening of the intensity of the beam and this weakening obeys an exponential law

$$I_x = I_0 e^{-sx},$$

where $s$ is the scattering attenuation coefficient or more briefly the scattering coefficient. The quotient $s/\rho$ of this number by the density of the scattering body is the mass scattering coefficient. If we call the atomic scattering coefficient $\sigma$ the ratio between the energy scattered in a single atom and the intensity of the incident radiation, we can easily see that it is related to $s$ by the equation:

$$\sigma = \frac{s}{\rho} A m_H.$$

$A$ is here the atomic weight of the scattering agent, $m_H$ the mass of the hydrogen atom. Substituting the numerical values in the factor

$$\frac{8\pi}{3}\frac{e^4}{m_0 c^4},$$

we find that

$$\sigma = 0.54 \times 10^{-24} p.$$

However, experience has shown that the $s/\rho$ ratio is very close to 0.2 so one should have:

$$\frac{A}{p} = \frac{0.54 \times 10^{-24}}{0.2 \times 1.46 \times 10^{-24}} = \frac{0.54}{0.29}.$$

This figure is close to 2, which is quite in line with our current understanding of the relationship between the number of intra-atomic electrons and atomic weight. Thomson's theory thus led to some interesting coincidences and the work of various experimentalists, including Barkla's, demonstrated, a long time ago, that it was broadly confirmed.[3]

## II. The Theory of Debye[4]

Difficulties remained. In particular, Dr. W. H. Bragg found in some cases a much stronger scattering than that accounted for in the previous theory and concluded that there was a proportionality of the scattered energy, not with the number of atomic electrons, but with the square of that number. Dr. Debye presented a more complete theory compatible with both Dr. Bragg's and Dr. Barkla's results.

Dr. Debye considers the intra-atomic electrons to be evenly distributed in a volume of about $10^{-8}$ cm; for ease of calculation, he even assumes that they are all distributed on the same circle. If the wavelength is large compared to the average distances between the electrons, their motions must be almost in phase and, in the total wave, the amplitudes radiated by each of them will be added. The scattered energy will then be proportional to $p^2$

[3] Past work on X-ray scattering is listed in the book by R. Ledoux-Lebard and A. Ledoux-Lebard. Daauvillier, *La physique des Rayons X.* Gauthier-Villars, 1921, beginning on p. 137.

[4] *Ann. d. Phys.*, 46, 1915, p. 809.

and no longer to $p$ so that the coefficient $\sigma$ will be written:

$$\sigma = \frac{8\pi}{3} \frac{e^4}{m_0^2 c^4} p^2.$$

As for distribution in space, it will be identical to the one that Mr. Thomson anticipated. For waves of progressively decreasing wavelengths, the distribution in space will become asymmetrical, the energy scattered in the direction from which the radiation comes from being much weaker than in the opposite direction. The reason for this is that one can no longer look at the vibrations of the various electrons as in phase when the wavelength becomes comparable to the mutual distances. The amplitudes radiated in the various directions will no longer add up because they are out of phase and the energy scattered will be less. However, in a cone of small aperture surrounding the extension of the direction of incidence, there will always be phase agreement and the amplitudes will be added; therefore for the directions contained in this cone, the scattering will be much greater than for the other directions. Dr. Debye predicted a curious phenomenon: as one progressively deviates from the cone axis defined above, the scattered intensity does not immediately decrease regularly, but first undergoes periodic variations; one should therefore observe on a screen placed perpendicularly to the transmitted beam clear and dark rings centered on the direction of the beam. Although Dr. Debye first thought that he recognized this phenomenon in some of Dr. Friedrich's experimental results, it does not seem to have been clearly observed so far.

For the short wavelengths, the phenomena must be simplified. The cone of strong scattering narrows more and more, the distribution becomes symmetrical again and must now comply with Thomson's formulas because the phases of the various electrons become completely incoherent, so it is the energies and not the amplitudes that are added.

The great interest of Dr. Debye's theory is to have explained the strong scattering of soft $X$-rays and to have shown how, when the frequency rises, the transition from this phenomenon to that of Thomson should take place. But it is essential to note that according to Debye's ideas, the higher the frequency, the more the symmetry of the scattered radiation and of the 0.2 value of coefficient $s/\rho$ must be effectively realized. However, we will see in the following paragraph that this is in no way the case.

## III. The Recent Theory of P. Debye and A. H. Compton[5]

Experiments in the field of hard $X$-rays and $\gamma$-rays have revealed facts that are very different from what previous theories can predict. First, as the frequency rises, the asymmetry of the scattered radiation increases; on the other hand, the total scattered energy decreases, the value of the mass coefficient $s/\rho$ tends to decrease rapidly as soon as the wavelength falls below 0.3 or 0.2 $A$ and becomes very low for $\gamma$. Thus, where Thomson's theory should apply better and better, it applies less and less.

Two other phenomena have been brought to light by recent experimental research, foremost among which are those of Mr. A. H. Compton. These have shown that scattering seems to be accompanied by a lowering of the frequency, which is variable with the direction of observation, and that it seems to cause motion of electrons. Almost simultaneously and independently, P. Debye and A. H. Compton managed to give an interpretation of these deviations from the classical laws based on the concept of the light quantum.

Here is the principle: if a quantum of light is deflected from its rectilinear path by passing in the vicinity of an electron, we must assume that during the time when the two energy cen-

[5] P. Debye, *Phys. Zeitschr.*, 24, 1923, 161-166; A. H. Compton, *Phys. Rev.*, 21, 1923, 207; 21 , 1923, 483; *Phil. Mag.*, 46, 1923,897.

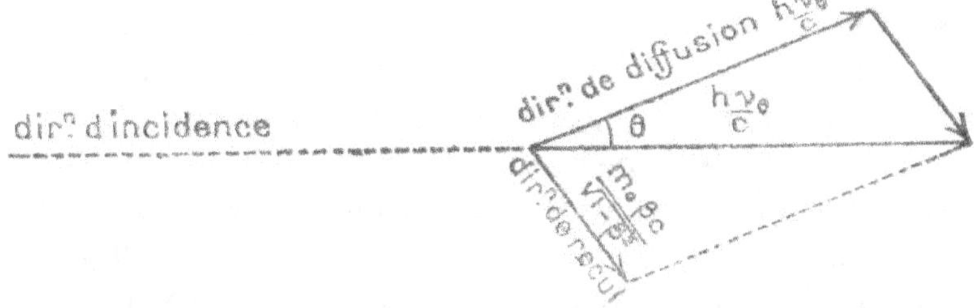


Fig. 6

ters are close enough, they exert a certain action on each other. When this action ends, the electron at rest will have borrowed some energy from the light corpuscle; according to the quantum relation, the scattered frequency will be less than the incident frequency. The conservation of momentum completes the problem. Suppose that the scattered quantum moves in a direction at the angle $\theta$ with the extension of the direction of incidence. The frequencies before and after the scattering being $\nu_0$ and $\nu_\theta$ and the electron's rest mass being $m_0$, we will have:

$$h\nu_\theta = h\nu_0 - m_0c^2\left(\frac{1}{\sqrt{1-\beta^2}} - 1\right)$$

$$\left(\frac{m_0\beta c}{\sqrt{I-\beta^2}}\right)^2 = \left(\frac{h\nu_0}{c}\right)^2 + \left(\frac{h\nu_\theta}{c}\right)^2 - 2\frac{h\nu_0}{c}\frac{h\nu_\theta}{c}\cos\theta.$$

This second relationship can be seen in the Figure 6. Velocity $v = \beta c$ is the velocity acquired by the electron by this process.

Let us designate by $\alpha$ the ratio $\frac{h\nu_0}{m_0c^2}$ equal to the quotient of $\nu_0$ by what we call the rest frequency of the electron. We obtain:

$$\nu_\theta = \frac{\nu_0}{1+2\alpha\sin^2\frac{\theta}{2}}$$

or

$$\lambda_\theta = \lambda_0\left(1+2\alpha\sin^2\frac{\theta}{2}\right).$$

One can also use these formulas to study the projection velocity and the direction of the "receding" electron. We find that the scattering directions varying from 0 to $\pi$, correspond for the electron to angles of recoil varying from $\pi/2$ to 0, the velocity varying simultaneously from 0 to a certain maximum.

Mr. Compton, using hypotheses inspired by the correspondence principle, thought he could calculate the value of the total scattered energy and thus explain the rapid decrease of the coefficient $s/\rho$. Mr. Debye applies the idea of correspondence in a slightly different form, but also manages to interpret the same phenomenon.

In an article in the *Physical Review* (May 1923), and in a more recent article in the *Philosophical Magazine* (November 1923), Mr. A. H. Compton showed that the new ideas outlined above accounted for many experimental facts and that, particularly for hard rays and light bodies, the expected wavelength variation was quantitatively verified. For heavier bodies and softer radiation, there seems to be a coexistence of a scattered line without frequency change and another scattered line according to Compton-Debye's law. For low frequencies, the first one becomes predominant and even seems to exist alone. Experiments by Dr. Ross on the scattering of the $MoK_\alpha$ line and green light by kerosene confirm this way of seeing. Line $K_\alpha$ gives a strong line scattered according to Compton's law and a weak line at unmodified frequency, the latter seems to exist alone for green light.

The existence of a non-displaced line seems to explain why crystalline reflection (the Laue phenomenon) is not accompanied by a variation in wavelength. Jauncey and Wolfers have recently shown that, if the lines scattered by the crystals usually used as reflectors, were appreciably affected by the Compton-Debye effect, precision measurements of Röntgen wavelengths would have already revealed the phenomenon. It must therefore be assumed that in this case, scattering takes place without quantum degradation.

At first glance, one is tempted to explain the existence of the two types of scattering in the following way: the Compton effect would occur whenever the scattering electron is free or at least when its bond with an atom corresponds to a low energy in comparison with that of the incident quantum; otherwise, there would be scattering without a change of wavelength because then the whole atom would take part in the process without acquiring any appreciable velocity because of its large mass. Dr. Compton finds it difficult to accept this idea and prefers to explain the unmodified line by the intervention of several electrons in the deviation of the same quantum; it would then be the high value of the sum of their masses that would prevent the passage of a notable energy from radiation to matter. In any case, it is easy to see why heavy elements and hard rays behave differently from light elements and soft rays.

As for how to render compatible the conception of scattering as the deviation of a luminous particle and the conservation of the phase necessary to explain the figures of Laue, it raises the considerable and still unresolved difficulties that we pointed out in the previous chapter with regard to Wave Optics.

When dealing with hard $X$-rays and light elements, as is the case in practice in Radiotherapy, the phenomena must be completely modified by the Compton effect and this is what seems to happen. We will give an example of this. It is known that in addition to weakening by scattering, an $X$-ray beam passing through matter experiences weakening by absorption, a phenomenon accompanied by the emission of photo-electrons. An empirical law due to Dr. Bragg and Dr. Pierce tells us that this absorption varies as the cube of the wavelength and undergoes abrupt discontinuities for all frequencies characteristic of the intra-atomic levels of the substance under consideration; moreover, for the same wavelength and various elements, the atomic coefficient of absorption varies as the fourth power of the atomic number.

This law is well verified in the mid-range Röntgen frequency domain and it seems likely that it should apply to hard rays. Since, according to the ideas accepted before Compton-Debye's theory, scattering was only a scattering of radiation, only the energy absorbed according to Bragg's law could produce ionization in a gas, the photoelectric electrons animated at high velocities ionizing by shocks the atoms encountered. The Bragg-Pierce law would thus allow to calculate the ionization ratio produced by the same hard radiation in two bulbs containing one a heavy gas (for example $CH^3I$) and the other a light gas (for example air). Even taking into account the numerous incidental corrections, this ratio was found experimentally much smaller than expected. Dr. Dauvillier had observed this phenomenon for $X$-rays and we were long intrigued by its interpretation.

The new scattering theory seems to effectively explain this anomaly. If, indeed, at least in the case of hard rays, part of the energy of the light quanta is transported to the scattered electron, there will be not only scattering of the radiation, but also "absorption by scattering". The ionization of the gas will be due both to the electrons expelled from the atom by the absorption mechanism itself and to the electrons set in motion by the scattering. In a heavy gas ($CH^3I$), Bragg absorption is intense and Compton absorption is almost negligible. For a light gas (air), it is no longer the case; the first absorption due to its variation in $N^4$ is very weak and the second, which is independent of $N$, becomes the most important. The ratio of the total absorptions and consequently of the ionizations in the two gases must therefore be much smaller than previously expected. It is even possible to give a quantitative account of the ratio of ionizations. This example shows that the new ideas of Compton and Debye are of great practical interest. The recoil of the scattered electrons seems to give the key to many other unexplained phenomena.

## IV. Scattering by Moving Electrons

The Compton-Debye theory can be generalized by considering the scattering of a quantum of radiation by a moving electron. Let us take as the $x$-axis the direction of primitive propagation of a quantum of initial frequency $\nu_1$, the $y$- and $z$-axes being arbitrarily chosen at right angles to each other in a plane normal to $ox$ and passing through the point where the scattering occurs. The direction of the electron velocity $\beta_1 c$ *before* the shock being defined by the director cosines $a_1 b_1 c_1$, we will call $\theta_1$ the angle it makes with $ox$, so that $\alpha_1 = \cos\theta_1$ ; after the shock, the scattered radiation quantum of frequency $\nu_2$ propagates in a direction of director cosines $pqr$ making the angle $\phi$ with the direction of the initial electron velocity ($\cos\phi = \alpha_1 p + b_1 q + c_1 r$) and the angle $\theta$ with the axis $ox$ ($p = \cos\theta$) . Finally the electron will have a final velocity $\beta_2 c$ whose director cosines will be $a_2 b_2 c_2$.

The conservation of, the energy and the amount of motion during the shock make it possible to write the following equations:

$$h\nu_1 + \frac{m_0 c^2}{\sqrt{1-\beta_1^2}} = h\nu_2 + \frac{m_0 c^2}{\sqrt{1-\beta_2^2}},$$

$$\frac{h\nu_1}{c} + \frac{m_0 \beta_1 c}{\sqrt{1-\beta_1^2}} a_1 = \frac{h\nu_2}{c} p + \frac{m_0 \beta_2 c}{\sqrt{1-\beta_2^2}} a_2,$$

$$\frac{m_0 \beta_1 c}{\sqrt{1-\beta_1^2}} b_1 = \frac{h\nu_2}{c} q + \frac{m_0 \beta_2 c}{\sqrt{1-\beta_2^2}} b_2,$$

$$\frac{m_0 \beta_1 c}{\sqrt{1-\beta_1^2}} c_1 = \frac{h\nu_2}{c} r + \frac{m_0 \beta_2 c}{\sqrt{1-\beta_2^2}} c_2.$$

Let's eliminate $a_2 b_2 c_2$ by means of the $a_2^2 + b_2^2 + c_2^2 = 1$ relationship; then, between the relationship thus obtained and the one that expresses energy conservation, let's eliminate $\beta_2$. Let's pose with Compton $\alpha = \frac{h\nu_1}{m_0 c^2}$ . We obtain:

$$\nu_2 = \nu_1 \frac{1 - \beta_1 \cos\theta_1}{1 - \beta_1 \cos\phi + 2\alpha\sqrt{1-\beta_1^2}\,\sin^2\frac{\theta}{2}}.$$

If the initial velocity of the electron is zero or negligible, we find

Compton's formula:

$$\nu_2 = \nu_1 \frac{1}{1 + 2\alpha \sin^2 \frac{\theta}{2}}.$$

In the general case, the Compton effect represented by the term in $\alpha$ remains but is diminished; moreover, a Doppler effect is added. If the Compton effect is negligible, one finds:

$$\nu_2 = \nu_1 \frac{1 - \beta_1 \cos\theta_1}{1 - \beta_1 \cos\phi}.$$

Since in this case, the scattering of the quantum does not interfere with the motion of the electron, we can expect to find a result identical to that of the electromagnetic theory. This is indeed what happens. Let us calculate the scattered frequency according to the electromagnetic theory (taking into account Relativity). The incident radiation has the frequency for the electron:

$$\nu' = \nu_1 \frac{1 - \beta_1 \cos\theta_1}{\sqrt{1 - \beta_1^2}}.$$

If the electron, while keeping translation velocity $\beta_1 c$, begins to vibrate at frequency $\nu'$, the observer who receives the radiation scattered in a direction making the angle $\phi$ with velocity $\beta_1 c$ of the source, assigns it frequency:

$$\nu_2 = \nu' \frac{\sqrt{1 - \beta_1^2}}{1 - \beta_1 \cos\phi}$$

we effectively have:

$$\nu_2 = \nu_1 \frac{1 - \beta_1 \cos\theta_1}{1 - \beta_1 \cos\phi}.$$

The Compton effect generally remains rather small, on the contrary, the Doppler effect can reach very high values for electrons accelerated by potential drops of a few hundred kilovolts (increase of one third of the frequency for 200 kilovolts). We are dealing here with a rise of the quantum because the scattering body, being animated with a high velocity, can yield energy to the radiation atom. The conditions for application of the Stokes rule are not realized. It is not impossible that some of the above conclusions could be subject to experimental verification at least with respect to $X$-rays.

# 7 Statistical Mechanics and Quanta

## I. Reminder of Some Results of Statistical Thermodynamics

The interpretation of the laws of thermodynamics using statistical considerations is one of the greatest achievements of scientific reasoning, but it is not without some difficulties and some objections. It is beyond the scope of the present work to make a critique of these methods; we will limit ourselves here, after recalling some fundamental results in their most widely used form today, to examine how our new ideas could be introduced into the theory of gases and black body radiation.

Boltzmann was the first to show that the entropy of a gas in a given state is, with the exception of an additive constant, the product of the logarithm of the probability of this state by the $k$ constant known as the "Boltzmann constant" which depends on the choice of the temperature scale; he first arrived at this conclusion by analyzing shocks between atoms under the hypothesis of a completely disordered agitation of the latter. Today, as a result of the work of Planck and Einstein, the relation $S = k \log P$ is considered as the definition of entropy $S$ of a system. In this definition, $P$ is not the mathematical probability equal to the quotient of the number of microscopic configurations giving the same total macroscopic configuration to the total number of possible configurations, it is the "thermodynamic" probability equal simply to the numerator of this fraction. This choice of meaning for $P$ amounts to determining (somehow arbitrarily) the entropy constant. Having accepted this postulate, we will recall a well-known demonstration of the analytical expression of thermodynamic quantities, a demonstration which has the advantage of being valid both when the sequence of possible states is discontinuous and in the opposite case.

Let's consider $\mathfrak{N}$ objects that can be arbitrarily distributed between m "states" or "cells" considered *a priori* as equally probable. A certain configuration of the system will be achieved by placing $n_1$ objects in cell 1, $n_2$ in cell 2, etc. The thermodynamic probability of this configuration will be:

$$P = \frac{\mathfrak{N}!}{n_1!\,n_2!\dots n_n!}.$$

If $\mathfrak{N}$ and all $n_i$ are large numbers, Stirling's formula gives for the entropy of the system:

$$S = k\log P = k\mathfrak{N}\log\mathfrak{N} - k\sum_1^m n_i \log n_i.$$

Let us suppose that to each cell, corresponds a given value of a certain function $\epsilon$ that we will name "the energy of an object placed in this cell". Let's consider a modification of the distribution of objects between cells subject to the condition that the sum of the energies remains invariable. The entropy $S$ will vary from:

$$dS = -k\delta\left[\sum_1^m n_i \log n_i\right] = -\kappa\sum_1^m \delta n_i - \kappa\sum_1^m \log n_i \delta n_i$$

with accompanying conditions:

$$\sum_1^m \delta n_i = 0 \qquad \text{and} \qquad \sum_1^m \epsilon_i \delta n_i = 0.$$

Maximum entropy is determined by the relation $\delta S = 0$. The method of indeterminate coefficients teaches us that, in order to realize this condition, the following equation must be satisfied:

$$\sum_{1}^{m} [\log ni + \eta + \beta\epsilon_i] \, \delta n_i = 0$$

where $\eta$ and $\beta$ are constants, and this is regardless of the $\delta n_i$.

This leads to the conclusion that the most likely distribution, which is the only one realized in practice, is governed by following law:

$$n_i = \alpha e^{-\beta\epsilon_i} \qquad \left(\alpha = e^{-\eta}\right).$$

This is the so-named "canonical" distribution. The thermodynamic entropy of the system corresponding to this most probable distribution is given by:

$$S = k\mathfrak{N} \log\mathfrak{N} - \sum_{1}^{m} \left[k\alpha \, e^{-\beta\epsilon_i} (\log\alpha - \beta\epsilon_i)\right]$$

or, since

$$\sum_{1}^{m} n_i = \mathfrak{N}$$

and

$$\sum_{1}^{m} \epsilon_i n_i = \text{the total energy } E$$

$$S = k\mathfrak{N} \log\frac{\mathfrak{N}}{\alpha} + k\beta E = k\mathfrak{N}\, log \sum_{1}^{m} e^{-\beta\epsilon_i} + k\beta E.$$

To determine $\beta$ we will use the thermodynamic relationship:

$$\begin{aligned}\frac{1}{T} + \frac{dS}{dE} &= \frac{\partial S}{\partial\beta}\frac{\partial\beta}{\partial E} + \frac{\partial S}{\partial E}\\ &= -k\mathfrak{N}\frac{\sum_1^m \epsilon_i e^{-\beta\epsilon_i}}{\sum_1^m e^{-\beta\epsilon_i}}\frac{d\beta}{dE} + kE\frac{d\beta}{dE} + k\beta\end{aligned}$$

and since

$$\mathfrak{N}\frac{\sum_1^m \epsilon_i e^{-\beta\epsilon_i}}{\sum_1^m e^{-\beta\epsilon_i}} = \mathfrak{N}\overline{\epsilon} = E$$

$$\frac{1}{T} = k\beta, \qquad \beta = \frac{1}{kT}.$$

Free energy is calculated by the relationship:

$$F = E - TS = E - k\mathfrak{N}\,T\log\left[\sum_1^m e^{-\beta\epsilon_i}\right] - \beta kTE$$

$$= -k\mathfrak{N}\,T\log\left[\sum_1^m e^{-\beta\epsilon_i}\right].$$

The average value of the free energy related to one of the objects is therefore:

$$\overline{F} = -kT\log\left[\sum_1^{\mathrm{m}} \mathrm{e}^{-\beta\epsilon_{\mathrm{i}}}\right].$$

Let's apply these general considerations to a gas formed of identical molecules of mass $m_0$. Liouville's theorem (also valid in the dynamics of relativity) tells us that the in-phase extension element of a molecule equal to $dxdydzdpdqdr$ (where $x, y$ and $z$ are the coordinates and $p, q, r$ are the corresponding moments) is an invariant of the equations of motion whose value is independent of the choice of coordinates. It was then assumed that the number of states of equal probability represented by an element of this phase extension is proportional to its size. This leads immediately to Maxwell's distribution law giving the number of atoms whose representative point (le point représentatif) falls within the element $dxdydzdpdqdr$:

$$dn = C^{te}\, e^{-\frac{w}{kT}}\, dxdydzdpdqdr,$$

where $w$ is the kinetic energy of these atoms.

Suppose the speeds are low enough to legitimize the use of classical dynamics, we then find:

$$w = \frac{1}{2}m_0 v^2 \qquad dpdqdr = 4\pi G^2 dG,$$

where $G = m_0 v = \sqrt{2m_0 w}$ is the momentum. Finally, the number of atoms contained in the volume element whose energy is between $w$ and $w + dw$ is given by the classical formula:

$$dn = C^{te}\, e^{-\frac{w}{kT}}\, 4\pi m_0^{\frac{3}{2}} \sqrt{2w}\, dw dx dy dz$$

The free energy and entropy remain to be calculated. For that purpose, we will take as object of the general theory, not an isolated molecule, but a whole gas formed of $N$ identical molecules of mass $m_0$ whose state is defined by $6N$ parameters. The free energy of the gas in the thermodynamic direction will be defined in the manner of Gibbs, as the mean value of the free energy of the $\mathfrak{N}$ gases, that is to say:

$$\overline{F} = -kT \log\left[\sum_1^m e^{-\beta\epsilon_i}\right] \qquad \beta = \frac{1}{kT}.$$

Mr. Planck specified how this sum should be calculated, it can be expressed as an integral extended to the entire $6N$-dimensional phase extension,[1] which is itself equivalent to the product of $N$ six-fold integrals extended to the phase extension of each molecule; but care must be taken to divide the result by $N!$, because of the identity of the molecules. The free energy being thus calculated, the entropy and the energy are deduced from it by the classical thermodynamic relations

$$S = -\frac{\partial F}{\partial T} \qquad E = F + TS.$$

To perform the calculations, the constant whose product by the extension element in phase gives the number of equally probable states represented by points of this element must be specified. This factor has the dimensions of the inverse of the cube of an action. Mr. Planck determines it from the following somewhat disconcerting hypothesis:

[1] EDITOR'S NOTE: In today's terms – space.

The phase extension of a molecule is divided into cells of equal probability whose value is finite and equal to $h^3$.

One can say either that within each cell there is a single point with a non-zero probability, or that all the points in the same cell correspond to physically indistinguishable states.

Planck's hypothesis leads to write for free energy:

$$F = -kT\log\left[\frac{1}{N!}\left(\int\int\int^{+}\int_{-\infty}^{\infty}\int\int e^{-\frac{\epsilon}{kT}}\frac{dxdydzdpdqdr}{h^3}\right)^N\right]$$
$$= -NkT\log\left[\frac{e}{N}\int\int\int^{+}\int_{-\infty}^{\infty}\int\int\frac{1}{h^3}e^{-\frac{\epsilon}{kT}}dxdydzdpdqdr\right].$$

One finds by performing the integration:

$$F = Nm_0c^2 - ?NT\log\left[\frac{eV}{Nh^3}(2\pi m_0 kT)^{\frac{3}{2}}\right]$$

$$V = \text{total gas volume}$$

$V$ = total gas volume and then,

$$S = kN\log\left[\frac{e^{5/2}V}{Nh^3}(2\pi m_0 kT)^{3/2}\right]$$

$$E = Nm_0c^2 + \frac{3}{2}kNT.$$

At the end of his book "Warmestrahlung" (4th ed.), Planck shows how the "chemical constant" involved in the equilibrium of a gas with its condensed phase can be deduced. Measurements of this chemical constant gave strong support to Planck's method.

So far we have involved neither Relativity nor our ideas about the relation of dynamics with the wave theory. We will try to find out how the previous formulas are modified by the introduction of these two notions.

## II. New Conception of the Statistical Equilibrium of a Gas

If the motion of the gaseous atoms is accompanied by wave propagation, the container containing the gas will be furrowed in all directions by these waves. We are naturally led to consider, as in the conception of black body radiation developed by Dr. Jeans, the phase waves forming stationary systems (i.e. resonating on the dimensions of the enclosure) as being the only stable ones; they alone would intervene in the study of the thermodynamic equilibrium. This is something similar to what we have encountered with the Bohr atom; there again, the stable trajectories were defined by a resonance condition and the others were to be considered as normally unachievable in the atom.

One could wonder how there can exist stationary phase wave systems in a gas since the motion of atoms is constantly disturbed by their mutual shocks. We can first answer that thanks to the uncoordinated molecular motion, the number of atoms diverted from their primitive direction during time $dt$ by the effect of the shocks is exactly compensated by the number of those whose motion is brought back in the same direction by the said effect; in short, everything happens as if the atoms follow a rectilinear trajectory from one wall to the other since their structural identity does not need to take into account their individuality. Moreover, during the duration of the free travel, the phase wave can travel several times the length of even a large container; if, for example, the average velocity of the atoms of a gas is $10^5$ cm/sec and the average travel is $10^{-5}$ cm, the average velocity of the phase waves will be $c^2/v = 9 \times 10^{15}$ cm/sec and during an average time of $10^{-10}$ seconds needed for the free travel, it will travel from $9 \times 10^5$ cm or 9 kilometers. It seems therefore possible to imagine the existence of standing phase waves in a gas mass in equilibrium.

To better understand the nature of the modifications that we will have to bring to statistical mechanics, we will first consider the simple case where molecules move along a straight line $AB$ of length $l$ while reflecting at $A$ and $B$. The initial distribution of positions and velocities is supposed to be defined by chance. The probability for a molecule to be on an element $dx$ of $AB$ is thus $dx/l$. From the classical conception, one must also take the probability of a velocity between $v$ and $v+dv$ proportional to $dv$; thus if one constitutes a phase extension by taking as variables $x$ and $v$, all the elements equal to $dxdv$ will be equally probable. The situation is quite different when the stability conditions considered above are introduced. If the velocities are low enough to neglect the terms of Relativity, the wavelength related to the motion of a molecule whose velocity is $v$ will be:

$$\lambda = \frac{\frac{c}{\beta}}{\frac{m_0 c^2}{h}} = \frac{h}{m_0 v}$$

and the resonance condition will be written as

$$l = n\lambda = n\frac{h}{m_0 v} \qquad (n \text{ is integer}).$$

Let us pose $h/m_0 l = v_0$, we obtain:

$$v = n v_0.$$

The velocity will therefore only take on values equal to integer multiples of $v_0$.

The variation $\delta n$ of integer $n$ corresponding to a variation $\delta v$ of the velocity gives the number of states of a molecule that are compatible with the existence of standing phase waves. We can see immediately that

$$\delta n = \frac{m_0 l}{h}\delta v.$$

So everything will happen as if each $\delta x \delta v$ element of the phase extension had $\frac{m_0}{h}\delta x \delta v$ possible states corresponding to it, which is the classical expression of the phase extension element divided by $h$. Examination of the numerical values shows that even an extremely small value of $\delta \upsilon$ for the scale of our experimental measurements corresponds to a large interval $\delta n$; any rectangle of the in-phase extension, even a very small one, corresponds to a huge number of "possible" values of $\upsilon$. In general, therefore, the quantity $\frac{m_0}{h}\delta x \delta \upsilon$ can be treated as a differential in the calculations. But, in principle, the distribution of representative points is no longer at all the one imagined by Statistical Mechanics; it is discontinuous and supposes that, by the action of a mechanism still impossible to specify, the motions of atoms that would be linked to non-stationary phase wave systems are automatically eliminated.

Let's now move on to the more concrete case of three-dimensional gas. The distribution of phase waves in the enclosure will be quite similar to that given by the old black body radiation theory for thermal waves. It will be possible, as Mr. Jeans did in this case, to calculate the number of standing waves contained in the unit of volume and whose frequencies are between $\nu$ and $\nu + \delta\nu$. We find for this number, by distinguishing group velocity $U$ from phase velocity $V$, the following expression:

$$n_\nu \delta\nu = \gamma \frac{4\pi}{UV^2} \nu^2 \delta\nu,$$

Where $\gamma$ is equal to 1 for longitudinal waves and 2 for transverse waves. The previous expression should not give us any illusion: not all the values of $\nu$ are present in the wave system and, if it is allowed to consider the above expression as a differential in calculations, it is because in general, in a very small frequency interval, there will be a huge number of admissible values for $\nu$.

Time has come to make use of the theorem demonstrated in Chapter I, paragraph II. To an atom of velocity $\upsilon = \beta c$, corresponds a wave having as phase velocity $V = c/\beta$, as group velocity $U = \beta c$ and as frequency

$$\nu = \frac{1}{h} \frac{m_0 c^2}{\sqrt{1-\beta^2}}.$$

If $w$ designates its kinetic energy, we find by the formulas of Relativity:

$$h\nu = \frac{m_0 c^2}{\sqrt{1-\beta^2}} = m_0 c^2 + w = m_0 c^2 (1+\alpha) \qquad \left(\alpha = \frac{w}{m_0 c^2}\right).$$

Whence:

$$n_w d_w = \gamma \frac{4\pi}{UV^2} \nu^2 d\nu = \gamma \frac{4\pi}{h^3} m_0^2 c(1+\alpha)\sqrt{\alpha(\alpha+2)} dw.$$

If we apply to the whole set of atoms the canonical distribution law shown above, we obtain for the number of those contained in the volume element $dxdydz$ and whose kinetic energy is between $w$ and $w + dw$:

$$C^{te} \gamma \frac{4\pi}{h^3} m_0^2 c(1+\alpha)\sqrt{\alpha(\alpha+2)} e^{-\frac{w}{kT}} dw dx dy dz. \qquad (1)$$

For material atoms, phase waves must be analogous to longitudinal waves for reasons of symmetry, so let's assume $\gamma = 1$. Moreover, for these atoms (except for a few atoms in negligible number at usual temperatures), the proper energy $m_0 c^2$ is infinitely higher than the kinetic energy. We can therefore equate $1+\alpha$ with the unit and find for the number defined above:

$$C^{te} \frac{4\pi}{h^3} m_0^{\frac{3}{2}} \sqrt{2w} e^{-\frac{w}{kT}} dw\, dx\, dy\, dz$$
$$= C^{te} e^{-\frac{w}{kT}} \int_w^{w+dw} \frac{dxdydzdpdqdr}{h^3}.$$

Obviously, our method leads us to take, in order to measure the number of possible states of the molecule corresponding to an element of its phase extension, not the quantity of this element itself but this quantity divided by $h^3$. We thus validate Dr.

Planck's hypothesis and, consequently, the results obtained by this researcher exposed above. It should be noted that it is the values found for the velocities $V$ and $U$ of the phase wave that have made it possible to arrive at this result from the Jeans formula.[2]

## III. The Gas of Light Atoms

If light is divided into atoms, the radiation of the black body can be considered as a gas of such atoms in equilibrium with matter a bit like a saturated vapor is in equilibrium with its condensed phase. We have already shown in Chapter III that this idea leads to an accurate prediction of the radiation pressure.

Let's try to apply to such a light gas the general Formula (1) of the previous paragraph. Here we must put $\gamma = 2$ because of the symmetry of the luminous unit which we insisted on in Chapter IV. Moreover, $\alpha$ is very large in relation to the unit, with the exception of a few atoms in negligible number at usual temperatures, which makes it possible to consider $\alpha + 1$ and $\alpha + 2$ as equal to $\alpha$. One would thus obtain for the number of atoms per volume element, energy between $h\nu$ and $h(\nu + d\nu)$:

$$C^{te}\frac{8\pi}{c^3}\nu^2 e^{-\frac{h\nu}{kT}}\,d\nu dx dy dz$$

and for the energy density corresponding to the same frequencies:

$$u_\nu d\nu = C^{te}\frac{8\pi h}{c^3}\nu^3 e^{-\frac{h\nu}{kT}}\,d\nu.$$

It would be easy to show that the constant is equal to $-1$ by following a reasoning contained in my article "Quanta de lu-

[2] On the issue covered in this paragraph, see : O. Sackur, *Ann. de Phys.*, 36, 958 (1911) and 40, 67 (1913); H. Tetrode, *Phys. Zeitschr.*, 14, 212 (1913); *Ann. de Phys.*, 38, 434 (1912); W. H. Keesom, *Phys. Zeitschr.*, 15, 695 (1914); O. Stern, *Phys. Zeitschr.*, 14, 629 (1913); E. Brody, *Zeitschr. f. Phys.*, 16, 79 (1921).

miére et rayonnement noir" published in the *Journal de Physique* in November 1922.

Unfortunately, the law thus obtained is Wien's law, which is only the first term in the series that constitutes Planck's experimentally exact law. This should not surprise us because, assuming the completely independent motions of light atoms, we must necessarily arrive at a law whose exponential factor is identical to that of Maxwell's law.

We also know that a continuous distribution of radiant energy in space would lead to Rayleigh's law as shown by Jeans' reasoning. However, Planck's law admits the expressions proposed by Wien and Lord Rayleigh as limit forms valid respectively for very large and very small values of quotient $\frac{h\nu}{kT}$. In order to find Planck's result, *a new hypothesis* must be made which, without departing from the conception of light quanta, allows us to explain how the classical formulae can be valid in a certain domain. We state this hypothesis as follows:

> If two or more atoms of light have exactly superimposed phase waves that can be said to be carried by the same wave, their motions can no longer be considered as completely independent and these atoms can no longer be treated as separate units in probability calculations.

The motion of these atoms "as a wave" would thus present a sort of coherence as a result of interactions that are impossible to specify, but are probably related to the mechanism that would make unstable the motion of atoms whose phase wave would not be stationary.

This hypothesis of coherence forces us to entirely revisit the demonstration of Maxwell's law. Since we can no longer take each atom of light as an "object" of the general theory, it is the elementary standing phase waves that must play this role. What do we call an elementary standing wave? A standing wave can be seen as the result of the superposition of two waves of formulas

$$\begin{matrix}\sin\\ \cos\end{matrix}\left[2\pi\left(\nu t - \frac{x}{\lambda} + \phi_0\right)\right]$$

and

$$\begin{matrix}\sin\\ \cos\end{matrix}\left[2\pi\left(\nu t + \frac{x}{\lambda} + \phi_0\right)\right]$$

where $\phi_0$ can take any value from 0 to 1. By giving $\nu$ one of the allowed values and $\phi_0$ an arbitrary value between 0 and 1, an elementary standing wave is defined. Let's consider a given value of $\phi_0$ and all allowed values of $\nu$ within a small range $d\nu$. Each elementary wave can carry 0, 1, 2... atoms and, since the canonical distribution law must be applicable to the considered waves, we find for the corresponding number of atoms:

$$N_\nu d\nu = n_\nu d_\nu \frac{\sum_1^\infty p e^{-p\frac{h\nu}{kT}}}{\sum_0^\infty e^{-p\frac{h\nu}{kT}}}.$$

By giving $\phi_0$ other values, other stable states will be obtained and by superimposing several of these stable states in such a way that the same standing wave corresponds to several elementary waves, a stable state will still be obtained. We conclude that the number of atoms whose total energy corresponds to frequencies between $\nu$ and $\nu + d\nu$ is

$$N_\nu d_\nu = A\gamma \frac{4\pi}{h^3} m_0^2 c(1+\alpha)\sqrt{\alpha(\alpha+2)}\, dw \sum_1^\infty e^{-p\frac{m_0c^2+w}{kT}}$$

per unit of volume. $A$ can be temperature dependent.

For a gas in the ordinary sense of the word, $m_0$ is so large that one can neglect all the terms of the series in front of the first one. So formula (1) of the previous paragraph is found.

For the luminous gas, we will now find:

$$N_\nu d\nu = A\frac{8\pi}{c^3}\nu^2 \sum_1^\infty e^{-p\frac{h\nu}{kT}} d\nu.$$

and, consequently, for the energy density:

$$u_\nu d\nu = A\frac{8\pi h}{c^3}\nu^3\sum_1^\infty e^{-p\frac{h\nu}{kT}}d\nu.$$

This is indeed the form of Planck's equation. But we must show that in this case $A = 1$. First of all, $A$ is here certainly a constant and not a function of temperature. Indeed, the total radiation energy per unit volume is:

$$u = \int_0^{+\infty} u_\nu d_\nu = A\frac{48\pi h}{c^3}\left(\frac{kT}{h}\right)^4\sum_1^\infty\frac{1}{p^4}.$$

and the total entropy is given by:

$$\begin{aligned} dS &= \frac{1}{T}[d(uV) + PdV] \\ &= V\frac{du}{T} + (u+P)\frac{dV}{T} \qquad \text{($V$ is the total volume)} \\ &= \frac{V}{T}\frac{du}{dT}dT + \frac{4}{3}u\frac{dV}{T} \end{aligned}$$

because

$$u = f(T) \qquad \text{and} \qquad P = \frac{1}{3}u - dS$$

is an exact differential, the integrability condition is written:

$$\frac{1}{T}\frac{du}{dT} = \frac{4}{3}\frac{1}{T}\frac{du}{dT} - \frac{4}{3}\frac{u}{T^2} \quad \text{or} \quad 4\frac{u}{T} = \frac{du}{dT} \qquad u = aT^4.$$

It is Stéfan's classical law that forces us to pose $A = C^{te}$. The preceding reasoning provides us with the values of entropy and free energy:

$$\begin{aligned} S &= A\frac{64\pi}{c^3h^3}k^4T^3V\sum_1^\infty\frac{1}{p^4} \\ F = U - TS &= -A\frac{16\pi}{c^3h^3}k^4T^4V\sum_1^\infty\frac{1}{p^4}. \end{aligned}$$

What remains to be determined is the value of the constant $A$. If we succeed in demonstrating that it corresponds to the unit, we will have recovered every formulas of Planck's theory.

As mentioned above, if we neglect the terms where $p > l1$, the demonstration is easy; the distribution of the atoms obeying the simple canonical law

$$A\frac{8\pi}{c^3}\nu^2 e^{-\frac{h\nu}{kT}}\,d\nu,$$

calculation of the free energy can be carried out with the Planck method in the same way as for an ordinary gas and, identifying the result with the above expression, we find $A = 1$.

In the general case, a more roundabout method must be used. Let us consider the $p^e$ term of the Planck series:

$$u_{\nu p}d\nu = A\frac{8\pi}{c^3}h\nu^3 e^{-p\frac{h\nu}{kT}}\,d\nu.$$

We can also write it as:

$$A\frac{8\pi}{c^3 p}\nu^2 e^{-p\frac{h\nu}{kT}}\,d\nu\, p\, h\nu$$

what allows asserting:

> The black body radiation can be considered as the mixture of an infinity of gases each characterized by an integer value $p$ and having the following property: the number of possible states of a gaseous unit located in an element of volume $dxdydz$ and having an energy between $ph\nu$ and $ph(\nu + d\nu)$ is equal to $\frac{8\pi}{c^3 p}\nu^2 d\nu dxdydz$.

Therefore, free energy can be calculated by the method of the first paragraph. The result is as follows:

$$\begin{aligned} F = \sum_1^\infty F_p &= -kT\sum_1^\infty \log\left[\frac{1}{n_p!}\left(V\int_0^\infty \frac{8\pi}{c^3 p}\nu^2 e^{-p\frac{h\nu}{kT}}\,d\nu\right)^{n_p}\right] \\ &= -kT\sum_1^\infty n_p \log\left[\frac{e}{n_p}V\int_0^{+\infty}\frac{8\pi}{c^3 p}\nu^2 e^{-p\frac{h\nu}{kT}}\,d_\nu\right] \end{aligned}$$

where

$$n_p = V\int_0^{+\infty} A\frac{8\pi}{pc^3}\nu^2 e^{-p\frac{h\nu}{kT}}\,d\nu = A\,\frac{16\pi}{c^3}\frac{k^3T^3}{h^3}\,\frac{1}{p^4}\,V.$$

Consequently:

$$F = -A\frac{16\pi}{c^3h^3}k^4T^4\log\left(\frac{e}{A}\right)\sum_1^\infty\frac{1}{p^4}\,V$$

and, by identification with the previously found expression:

$$\log\left(\frac{e}{A}\right) = 1 \qquad A = 1.$$

Which is what we meant to demonstrate.

The hypothesis of coherence adopted above thus led us to a successful conclusion by avoiding contradicting Rayleigh's law or Wien's law. A study of the fluctuations of black body radiation will provide us with a further proof of its importance.

## IV. Energy Fluctuations in the Black Body Radiation[3]

If grains of energy of value $q$ are distributed in very large numbers in a certain space and if their positions vary continuously according to the laws of chance, an element of volume will normally contain $\overline{n}$ grains, i.e. $\overline{E} = \overline{n}q$ energy. But the real value of $n$ will constantly deviate from $\overline{n}$ and one will have $\overline{(n-\overline{n}))^2} = \overline{n}$ according to a known theorem of probability theory and, consequently, the root mean square fluctuation of the energy will be:

$$\overline{\epsilon^2} = \overline{(n-\overline{n})^2}q^2 = \overline{n}q^2 = \overline{E}q.$$

[3] *La théorie du Rayonnement noir et les quanta,* Solvay meating, report of Mr. Einstein, p. 419; *Les théories statistiques en thermodynarnique,* Conferences by Mr. H.-A. Lorentz at the Collége de France, Teubner, 1916, pp. 70 et 114.

On the other hand, it is known that energy fluctuations in a volume $V$ of black body radiation are governed by the law of statistical thermodynamics:

$$\overline{\epsilon^2} = kT^2 V \frac{d(u_\nu d\nu)}{dT}$$

as long as they refer to the frequency interval $\nu, \nu + d\nu$. If one accepts Rayleigh's law:

$$u_\nu = \frac{8\pi k}{c^3}\nu^2 T, \qquad \overline{\epsilon^2} = \frac{c^3}{8\pi\nu^3 d\nu}\,\frac{(Vu_\nu d\nu)^2}{V}$$

and this result, as was to be expected, coincides with that provided by interference calculations carried out according to the rules of electromagnetic theory. If, on the contrary, we adopt Wien's law, which corresponds to the hypothesis of a radiation formed of completely independent atoms, we find:

$$\overline{\epsilon^2} = kT^2 V \frac{d}{dT}\left(\frac{8\pi h}{c^3}\nu^3 e^{-\frac{h\nu}{kT}} d\nu\right) = (u_\nu V d\nu)h\nu,$$

a formula which is also derived from $\overline{\epsilon^2} = \overline{E}h\nu$.

Finally, in the real case of Planck's law, as Einstein first noticed, we end up with expression:

$$\overline{\epsilon^2} = (u_\nu V d\nu)h\nu + \frac{c^3}{8\pi\nu^2 d\nu}\,\frac{(u_\nu d_\nu V)^2}{V}$$

$\overline{\epsilon^2}$ thus appears as the sum of what it would be:

1° if the radiation was formed of independent $h\nu$ quanta;

2° if the radiation was purely wave-like.

On the other hand, the conception of groups of atoms "as waves" leads to writing Planck's law as:

$$u_\nu d\nu = \sum_1^\infty \frac{8\pi h}{c^3}\nu^3 e^{-p\frac{h\nu}{kT}} d\nu = \sum_1^\infty n_{p,\nu} p h\nu\, d\nu$$

and, applying the formula $\overline{\epsilon^2} = \overline{n}q^2$ to each sort of grouping, one obtains:

$$\overline{\epsilon^2} = \sum_1^\infty n_{p,\nu} d\nu (ph\nu)^2.$$

Of course this expression is basically the same as Einstein's; only the way of writing is different. But its interest lies in leading us to the following statement:

> One can also correctly evaluate the fluctuations of the black body radiation not by using any interference theory, but by introducing the coherence of the atoms linked to the same phase wave.

It therefore seems almost certain that any attempt to reconcile the discontinuity of radiant energy and interferences should involve the coherence hypothesis of the last paragraph.

# APPENDIX TO CHAPTER 5
# ON LIGHT QUANTA

We proposed to consider light atoms as small energy centers characterized by a very low proper mass $m_0$ and animated with a velocity generally very close to $c$, so that there exists between the frequency $\nu$, the proper mass $m_0$ and the velocity $\beta c$ the relation:

$$h\nu = \frac{m_0 c^2}{\sqrt{1-\beta^2}},$$

from which we deduce:

$$\beta = \sqrt{1 - \left(\frac{m_0 c^2}{h\nu}\right)^2}.$$

This way of seeing things led us to remarkable concordances regarding the Doppler effect and radiation pressure.

Unfortunately, it raises an important difficulty: for lower and lower frequencies $\nu$, the velocity $\beta c$ of the radiant energy would become smaller and smaller, would cancel for $h\nu = m_0 c^2$ and then become imaginary (?).[1] This is all the more difficult to admit since, in the very low frequencies domain, one should expect to find the conclusions of the old theories that assign velocity c to the radiant energy.

This objection is very interesting because it draws attention to the transition from the purely corpuscular form of light, which

[1] EDITOR'S NOTE: The question mark is by de Broglie.

manifests itself in the high frequency range, to the purely undulatory form of very low frequencies. We have shown in Chapter VII that the purely corpuscular concept leads to Wien's law while, as is well known, the purely wave-like concept leads to Rayleigh's law. The passage from one to the other of these laws must, it seems to me, be closely related to the answers that may be given to the objection stated above.

I will, rather as an example than in the hope of providing a satisfactory solution, develop an idea suggested by the preceding reflections.

In Chapter VII, I showed that it was possible to interpret the transition from Wien's law to Rayleigh's law by conceiving of the existence of sets of atoms of light linked to the propagation of the *same* phase wave. I insisted on the similarity that such a wave carrying many quanta will bear with the classical wave when the number of quanta increases indefinitely. However, this similarity would be limited in the concept outlined in the text by the fact that each energy grain would keep its own very small, but finite proper mass $m_0$, whereas electromagnetic theory assigns to light a null mass. The frequency of the wave with multiple energy centers is determined by relationship:

$$h\nu = \frac{\mu_0 c^2}{\sqrt{1-\beta^2}},$$

where $\mu_0$ is the proper mass of *each* center: this seems required to account for the emission and absorption of energy in finite quantities $h\nu$. But we could perhaps suppose that the mass of the energy centers related to the *same wave* differs from the proper mass $m_0$ of an isolated center and depends on the number of other centers with which they interact. We would then have:

$$\mu_0 = f(p) \qquad \text{with} \qquad f(1) = m_0$$

by designating with $p$ the number of centers carried by the wave.

The need to fall back to the formulas of electromagnetism for very low frequencies, would lead us to suppose that $f(p)$ is a decreasing function of $p$ tending towards 0 when $p$ tends towards infinity. The velocity of all $p$ centers forming a wave would then be:

$$\beta c = c\sqrt{1 - \left[\frac{f(p)c^2}{h\nu}\right]^2}.$$

For very high frequencies, $p$ would almost always be equal to 1, the energy grains would be isolated, we would have Wien's law for the black body radiation and the formula

$$\beta = \sqrt{1 - \frac{m_0^2 c^4}{h^2 \nu^2}}$$

for the velocity of radiant energy.

For very low frequencies, $p$ would always be very large, the grains would be gathered in very numerous groups on the same wave. The radiation of the black body would obey Rayleigh's law and the velocity would tend towards $c$ when $\nu$ would tend towards 0.

The previous hypothesis destroys a little the simplicity of the concept of the "quantum of light", but this simplicity can certainly not be entirely preserved if one wants to be able to connect the electromagnetic theory with the discontinuity revealed by the photoelectric phenomena. This connection would be obtained, it seems to me, by the introduction of the function $f(p)$ because, for a given energy, a wave will have to include an increasingly larger number $p$ of grains when $\nu$ and $h\nu$ decrease; when the frequency becomes lower and lower, the number of grains must increase indefinitely, their proper mass $m_0$ tending towards 0 and their speed towards $c$, so that the wave carrying them would become more and more analogous to the electromagnetic wave.

It must be admitted that the real structure of light energy is still very mysterious.

## Summary and Conclusions

In a rapid historical review of the development of Physics since the 17th century, and in particular of Dynamics and Optics, we have shown how the issue of quanta was somehow contained in the parallelism of corpuscular and undulatory conceptions of radiation; then, we have reminded how intensely the notion of quanta imposed itself to the attention of 20th century scientists. In the first chapter, we admitted as a fundamental postulate the existence of a periodic phenomenon linked to each isolated piece of energy and dependent on its proper mass by means of the Planck-Einstein relationship. The Relativity Theory then showed us the need to associate to the uniform motion of any moving entity the propagation at constant velocity of a certain "phase wave" and we were able to interpret this propagation by considering Minkowski's space-time.

Taking up, in Chapter II, the same issue in the more general case of an electrically charged body moving with a non-uniform motion in an electromagnetic field, we showed that, according to our ideas, the principle of least action in its Maupertuisian form and the principle of phase concordance due to Fermat could well be two aspects of a single law; this led us to conceive an extension of the quantum relation giving the velocity of the phase wave in the electromagnetic field. Certainly, this idea that the motion of a material point always conceals the propagation of a wave would need to be studied and completed, but if it could be given a fully satisfactory form, it would represent a synthesis of great rational beauty.

The main consequence that can be drawn from it is discussed in Chapter III. After recalling the stability laws of quantized trajectories as they emerge from numerous recent works, we have shown that they can be interpreted as expressing the resonance of the phase wave over the length of closed or quasi-closed trajectories. We believe that this is the first physically plausible explanation proposed for these Bohr-Sommerfeld stability conditions.

The difficulties raised by the simultaneous motion of two electrical centers are studied in Chapter IV, particularly in the case of circular motions of the nucleus and the electron about their common center of gravity in the hydrogen atom.

In Chapter V, guided by the results previously obtained, we try to imagine the possibility of a concentration of radiant energy around certain singular points and we show the deep harmony that seems to exist between the opposing viewpoints of Newton and Fresnel and which is revealed by the identity of many predictions. The electromagnetic theory cannot be fully preserved in its present form, but its reworking is a difficult task, and we suggest a qualitative theory of interference.

In Chapter VI, we summarize the various successive theories of scattering of $X$ and $\gamma$ rays by amorphous bodies, with particular emphasis on the very recent theory of P. Debye and A. H. Compton, that apparently render almost tangible the existence of light quanta.

Finally, in chapter VII, we introduce the phase wave in Statistical Mechanics. We also find the value of the phase extension element proposed by Planck and we obtain the law of radiation of the black body as well as Maxwell's law of a gas formed of atoms of light on the condition however that we admit a certain coherence between the motions of certain atoms, a coherence whose study of energy fluctuations also seems to be of interest.

In short, I developed new ideas that could perhaps contribute to hastening the necessary synthesis that will once again unify radiation physics, today so strangely divided into two domains in which two opposing conceptions reign: the corpuscular and the wave conception. I sensed that the principles of the material point Dynamics, if we knew how to analyze them correctly, would undoubtedly present themselves as expressing both propagations and phase concordances, and I tried my best to draw from them the explanation of a certain number of enigmas posed by the Quanta theory. In attempting to do so, I came to some interesting conclusions that perhaps give hope for more complete results by continuing along the same path. But first of all, a new electromagnetic theory would have to be constituted, naturally conforming to the principle of Relativity, taking into account the discontinuous structure of radiant energy and the physical nature of phase waves, finally leaving to Maxwell-Lorentz's theory a character of statistical approximation which would explain the legitimacy of its use and the accuracy of its predictions in a very large number of cases.

I intentionally left rather vague the definitions of the phase wave and of the periodic phenomenon of which it would be a sort of expression as well as that of the quantum of light. The present theory must therefore be considered more as a form whose physical content is not entirely specified than as a homogeneous doctrine definitively constituted.

# The Wave Nature of the Electron Louis de Broglie's Nobel Lecture, December 12, 1929[1]

When in 1920 I resumed my studies of theoretical physics which had long been interrupted by circumstances beyond my control, I was far from the idea that my studies would bring me several years later to receive such a high and envied prize as that awarded by the Swedish Academy of Sciences each year to a scientist: the Nobel Prize for Physics. What at that time drew me towards theoretical physics was not the hope that such a high distinction would ever crown my work; I was attracted to theoretical physics by the mystery enshrouding the structure of matter and the structure of radiations, a mystery which deepened as the strange quantum concept introduced by Planck in 1900 in his research on black-body radiation continued to encroach on the whole domain of physics.

To assist you to understand how my studies developed, I must first depict for you the crisis which physics had then been passing through for some twenty years.

For a long time physicists had been wondering whether light was composed of small, rapidly moving corpuscles. This idea was put forward by the philosophers of antiquity and upheld

[1]Louis de Broglie – Nobel Lecture. NobelPrize.org. Nobel Media AB 2021. Sun. 17 Jan 2021. https://www.nobelprize.org/prizes/physics/1929/broglie/lecture/

by Newton in the 18th century. After Thomas Young's discovery of interference phenomena and following the admirable work of Augustin Fresnel, the hypothesis of a granular structure of light was entirely abandoned and the wave theory unanimously adopted. Thus the physicists of last century spurned absolutely the idea of an atomic structure of light. Although rejected by optics, the atomic theories began making great headway not only in chemistry, where they provided a simple interpretation of the laws of definite proportions, but also in the physics of matter where they made possible an interpretation of a large number of properties of solids, liquids, and gases. In particular they were instrumental in the elaboration of that admirable kinetic theory of gases which, generalized under the name of statistical mechanics, enables a clear meaning to be given to the abstract concepts of thermodynamics. Experiment also yielded decisive proof in favour of an atomic constitution of electricity; the concept of the electricity corpuscle owes its appearance to Sir J. J. Thomson and you will all be familiar with H. A. Lorentz's use of it in his theory of electrons.

Some thirty years ago, physics was hence divided into two: firstly the physics of matter based on the concept of corpuscles and atoms which were supposed to obey Newton's classical laws of mechanics, and secondly radiation physics based on the concept of wave propagation in a hypothetical continuous medium, i.e. the light ether or electromagnetic ether. But these two physics could not remain alien one to the other; they had to be fused together by devising a theory to explain the energy exchanges between matter and radiation – and that is where the difficulties arose. While seeking to link these two physics together, imprecise and even inadmissible conclusions were in fact arrived at in respect of the energy equilibrium between matter and radiation in a thermally insulated medium: matter, it came to be said, must yield all its energy to the radiation and so tend of its own accord to absolute zero temperature! This absurd conclusion had at all costs to be avoided. By an intuition of his genius Planck realized the way of avoiding it: instead of assuming, in common with the classical wave theory, that a light source emits its radiation continuously, it had to be assumed on the contrary that it emits equal and finite quantities, quanta. The energy of each quantum has, moreover, a value proportional to the frequency $\nu$ of the radiation. It is equal to $h\nu$, $h$ being a universal constant since referred to as Planck's constant.

The success of Planck's ideas entailed serious consequences. If light is emitted as quanta, ought it not, once emitted, to have a granular structure? The existence of radiation quanta thus implies the corpuscular concept of light. On the other hand, as shown by Jeans and H. Poincaré, it is demonstrable that if the motion of the material particles in light sources obeyed the laws of classical mechanics it would be impossible to derive the exact law of black-body radiation, Planck's law. It must therefore be assumed that traditional dynamics, even as modified by Einstein's theory of relativity, is incapable of accounting for motion on a very small scale.

The existence of a granular structure of light and of other radiations was confirmed by the discovery of the photoelectric effect. If a beam of light or of $X$-rays falls on a piece of matter, the latter will emit rapidly moving electrons. The kinetic energy of these electrons increases linearly with the frequency of the incident radiation and is independent of its intensity. This phenomenon can be explained simply by assuming that the radiation is composed of quanta $h\nu$ capable of yielding all their energy to an electron of the irradiated body: one is thus led to the theory of light quanta proposed by Einstein in 1905 and which is, after all, a reversion to Newton's corpuscular theory, completed by the relation for the proportionality between the energy of the corpuscles and the frequency. A number of arguments were put forward by Einstein in support of his viewpoint and in 1922 the discovery by A. H. Compton of the $X$-ray scattering phenomenon which bears his name confirmed it. Nevertheless, it was still necessary to adopt the wave theory to account for interference and diffraction phenomena and no way whatsoever of reconciling the wave theory with the existence of light corpuscles could be visualized.

As stated, Planck's investigations cast doubts on the validity of very small scale mechanics. Let us consider a material point which describes a small trajectory which is closed or else turning back on itself. According to classical dynamics there are numberless motions of this type which are possible complying with the initial conditions, and the possible values for the energy of the moving body form a continuous sequence. On the other hand Planck was led to assume that only certain preferred motions, quantized motions, are possible or at least stable, since energy can only assume values forming a discontinuous sequence. This concept seemed rather strange at first but its value had to be recognized because it was this concept which brought Planck to the correct law of black-body radiation and because it then proved its fruitfulness in many other fields. Lastly, it was on the concept of atomic motion quantization that Bohr based his famous theory of the atom; it is SO familiar to scientists that I shall not summarize it here.

The necessity of assuming for light two contradictory theories-that of waves and that of corpuscles – and the inability to understand why, among the infinity of motions which an electron ought to be able to have in the atom according to classical concepts, only certain ones were possible: such were the enigmas confronting physicists at the time I resumed my studies of theoretical physics.

When I started to ponder these difficulties two things struck me in the main. Firstly the light-quantum theory cannot be regarded as satisfactory since it defines the energy of a light corpuscle by the relation $W = h\nu$ which contains a frequency $\nu$. Now a purely corpuscular theory does not contain any element permitting the definition of a frequency. This reason alone renders it necessary in the case of light to introduce simultaneously the corpuscle concept and the concept of periodicity.

On the other hand the determination of the stable motions of the electrons in the atom involves whole numbers, and so far the only phenomena in which whole numbers were involved in physics were those of interference and of eigenvibrations. That suggested the idea to me that electrons themselves could not be represented as simple corpuscles either, but that a periodicity had also to be assigned to them too.

I thus arrived at the following overall concept which guided my studies: for both matter and radiations, light in particular, it is necessary to introduce the corpuscle concept and the wave concept at the same time. In other words the existence of corpuscles accompanied by waves has to be assumed in all cases. However, since corpuscles and waves cannot be independent because, according to Bohr's expression, they constitute two complementary forces of reality, it must be possible to establish a certain parallelism between the motion of a corpuscle and the propagation of the associated wave. The first objective to achieve had, therefore, to be to establish this correspondence.

With that in view I started by considering the simplest case: that of an isolated corpuscle, i.e. a corpuscle free from all outside influence. We wish to associate a wave with it. Let us consider first of all a reference system $O, x_0, y_0, z_0$ in which the corpuscle is immobile: this is the "intrinsic" system of the corpuscle in the sense of the relativity theory. In this system the wave will be stationary since the corpuscle is immobile: its phase will be the same at every point; it will be represented by an expression of the form $\sin 2\pi\nu_0(t_0 - \tau_0)$; $t_0$ being the intrinsic time of the corpuscle and $\tau_0$ a constant.

In accordance with the principle of inertia in every Galilean system, the corpuscle will have a rectilinear and uniform motion. Let us consider such a Galilean system and let $v = \beta c$ be the velocity of the corpuscle in this system; we shall not restrict generality by taking the direction of the motion as the $x$-axis. In compliance with Lorentz' transformation, the time $t$ used by an observer of this new system will be associated with the intrinsic time $t_0$ by the relation:

$$t_0 = \frac{t - \frac{\beta x}{c}}{\sqrt{1-\beta^2}}$$

and hence for this observer the phase of the wave will be given by

$$\sin 2\pi \frac{\nu_0}{\sqrt{1-\beta^2}} \left( t - \frac{\beta x}{c} - \tau_0 \right).$$

For him the wave will thus have a frequency:

$$\nu = \frac{\nu_0}{\sqrt{1-\beta^2}}$$

and will propagate in the direction of the $x$-axis at the phase velocity:

$$V = \frac{c}{\beta} = \frac{c^2}{v}.$$

By the elimination of $\beta$ between the two preceding formulae the following relation can readily be derived which defines the refractive index of the vacuum $n$ for the waves considered:

$$n = \sqrt{1 - \frac{\nu_0}{\nu^2}}.$$

A "group velocity" corresponds to this "law of dispersion." You will be aware that the group velocity is the velocity of the resultant amplitude of a group of waves of very close frequencies. Lord Rayleigh showed that this velocity $U$ satisfies equation :

$$\frac{1}{U} = \frac{1}{c} \frac{\partial (n\nu)}{\partial \nu}.$$

Here $U = v$, that is to say that the group velocity of the waves in the system $x, y, z, t$ is equal to the velocity of the corpuscle in this system. This relation is of very great importance for the development of the theory.

The corpuscle is thus defined in the system $x, y, z, t$ by the frequency $\nu$ and the phase velocity $V$ of its associated wave. To establish the parallelism of which we have spoken, we must seek to link these parameters to the mechanical parameters, energy and quantity of motion. Since the proportionality between energy and frequency is one of the most characteristic relations of the quantum theory, and since, moreover, the frequency and the energy transform in the same way when the Galilean reference system is changed, we may simply write

$$\text{energy} = h \times \text{frequency} \quad \text{or} \quad W = h\nu,$$

where $h$ is Planck's constant. This relation must apply in all Galilean systems and in the intrinsic system of the corpuscle where the energy of the corpuscle, according to Einstein, reduces to its internal energy $m_0c^2$ ($m_0$ being the rest mass) we have

$$h\nu_0 = m_0c^2.$$

This relation defines the frequency $\nu_0$ as a function of the rest mass $m_0$, or inversely.

The quantity of movement is a vector $p$ equal to

$$\frac{m_0 v}{\sqrt{1-\beta^2}}$$

and we have:

$$(p) = \frac{m_0 v}{\sqrt{1-\beta^2}} = \frac{Wv}{c^2} = \frac{h\nu}{V} = \frac{h}{\lambda}.$$

The quantity $\lambda$ is the distance between two consecutive peaks of the wave, i.e. the "wavelength." Hence:

$$\lambda = \frac{h}{p}.$$

This is a fundamental relation of the theory.

The whole of the foregoing relates to the very simple case where there is no field of force at all acting on the corpuscles. I shall show you very briefly how to generalize the theory in the case of a corpuscle moving in a constant field of force deriving from a potential function $F(xyz)$. By reasoning which I shall pass over, we are then led to assume that the propagation of the wave corresponds to a refractive index which varies from point to point in space in accordance with the formula:

$$n(xyz) = \sqrt{\left[1 - \frac{F(xyz)}{h\nu}\right]^2 - \frac{\nu_0}{\nu^2}}$$

or to a first approximation if the corrections introduced by the theory of relativity are negligible

$$n(xyz) = \sqrt{\frac{2(E-F)}{m_0 c^2}}$$

with

$$E = W - m_0 c^2.$$

The constant energy $W$ of the corpuscle is still associated with the constant frequency $\nu$ of the wave by the relation

$$W = h\nu,$$

while the wavelength $\lambda$ which varies from one point to another of the force field is associated with the equally variable quantity of motion $p$ by the following relation

$$\lambda(xyz) = \frac{h}{p(xyz)}.$$

Here again it is demonstrated that the group velocity of the waves is equal to the velocity of the corpuscle. The parallelism thus established between the corpuscle and its wave enables us to identify Fermat's principle for the waves and the principle of least action for the corpuscles (constant fields). Fermat's principle states that the ray in the optical sense which passes through two points $A$ and $B$ in a medium having an index $n(xyz)$ varying from one point to another but constant in time is such that the integral

$$\int_A^B n\,dl$$

taken along this ray is extreme. On the other hand Maupertuis' principle of least action teaches us the following: the trajectory of a corpuscle passing through two points $A$ and $B$ in space is such that the integral

$$\int_A^B p\,dl$$

taken along the trajectory is extreme, provided, of course, that only the motions corresponding to a given energy value are considered. From the relations derived above between the mechanical and the wave parameters, we have:

$$n = \frac{c}{V} = \frac{c}{\nu}\frac{1}{\lambda} = \frac{c}{h\nu}\frac{h}{\lambda} = \frac{c}{W}p = \text{const.}\, p$$

since $W$ is constant in a constant field. It follows that Fermat's and Maupertuis' principles are each a translation of the other and the possible trajectories of the corpuscle are identical to the possible rays of its wave.

These concepts lead to an interpretation of the conditions of stability introduced by the quantum theory. Actually, if we consider a closed trajectory $C$ in a constant field, it is very natural to assume that the phase of the associated wave must be a uniform function along this trajectory. Hence we may write :

$$\int_c \frac{dl}{\lambda} = \int_c \frac{1}{h} p\,dl = \text{integer.}$$

This is precisely Planck's condition of stability for periodic atomic motions. The conditions of quantum stability thus emerge as analogous to resonance phenomena and the appearance of integers becomes as natural here as in the theory of vibrating cords and plates.

The general formulae which establish the parallelism between waves and corpuscles may be applied to corpuscles of light on the assumption that here the rest mass $m_0$ is infinitely small. Actually, if for a given value of the energy $W$, $m_0$ is made to tend towards zero, $v$ and $V$ are both found to tend towards $c$ and at the limit the two fundamental formulae are obtained on which Einstein had based his light-quantum theory

$$W = h\nu \qquad p = \frac{h\nu}{c}.$$

Such are the main ideas which I developed in my initial studies. They showed clearly that it was possible to establish a correspondence between waves and corpuscles such that the laws of mechanics correspond to the laws of geometrical optics. In the wave theory, however, as you will know, geometrical optics is only an approximation: this approximation has its limits of validity and particularly when interference and diffraction phenomena are involved, it is quite inadequate. This prompted the thought that classical mechanics is also only an approximation relative to a vaster wave mechanics. I stated as much almost at the outset of my studies, i.e. "A new mechanics must be developed which is to classical mechanics what wave optics is to geometrical optics." This new mechanics has since been developed, thanks mainly to the fine work done by Schroödinger. It is based on wave propagation equations and strictly defines the evolution in time of the wave associated with a corpuscle. It has in particular succeeded in giving a new and more satisfactory form to the quantization conditions of intra-atomic motion since the classical quantization conditions are justified, as we have seen, by the application of geometrical optics to the waves associated with the intra-atomic corpuscles, and this application is not strictly justified.

I cannot attempt even briefly to sum up here the development of the new mechanics. I merely wish to say that on examination it proved to be identical with a mechanics independently developed, first by Heisenberg, then by Born, Jordan, Pauli, Dirac, etc.: quantum mechanics. The two mechanics, wave and quantum, are equivalent from the mathematical point of view.

We shall content ourselves here by considering the general significance of the results obtained. To sum up the meaning of wave mechanics it can be stated that: "A wave must be associated with each corpuscle and only the study of the wave's propagation will yield information to us on the successive positions of the corpuscle in space." In conventional large-scale mechanical phenomena the anticipated positions lie along a curve which is the trajectory in the conventional meaning of the word. But what happens if the wave does not propagate according to the laws of optical geometry, if, say, there are interferences and diffraction? Then it is no longer possible to assign to the corpuscle a motion complying with classical dynamics, that much is certain. Is it even still possible to assume that at each moment the corpuscle occupies a well-defined position in the wave and that the wave in its propagation carries the corpuscle along in the same way as a wave would carry along a cork? These are difficult questions and to discuss them would take us too far and even to the confines of philosophy. All that I shall say about them here is that nowadays the tendency in general is to assume that it is not constantly possible to assign to the corpuscle a well-defined position in the wave. I must restrict myself to the assertion that when an observation is carried out enabling the localization of the corpuscle, the observer is invariably induced to assign to the corpuscle a position in the interior of the wave and the probability of it being at a particular point $M$ of the wave is proportional to the square of the amplitude, that is to say the intensity at $M$.

This may be expressed in the following manner. If we consider a cloud of corpuscles associated with the same wave, the intensity of the wave at each point is proportional to the cloud density at that point (i.e. to the number of corpuscles per unit volume around that point). This hypothesis is necessary to explain how, in the case of light interferences, the light energy is concentrated at the points where the wave intensity is maximum: if in fact it is assumed that the light energy is carried by light corpuscles, photons, then the photon density in the wave must be proportional to the intensity.

This rule in itself will enable us to understand how it was possible to verify the wave theory of the electron by experiment.

Let us in fact imagine an indefinite cloud of electrons all moving at the same velocity in the same direction. In conformity with the fundamental ideas of wave mechanics we must associate with this cloud an indefinite plane wave of the form

$$a \sin 2\pi \left[ \frac{W}{h} t - \frac{\alpha x + \beta y + \gamma z}{\lambda} \right],$$

where $\alpha\beta\gamma$ are the cosines governing the propagation direction and where the wavelength $\lambda$ is equal to $h/p$. With electrons which are not extremely fast, we may write

$$p = m_0 v$$

and hence

$$\lambda = \frac{h}{m_0 v}$$

where $m_0$ is the rest mass of the electron.

You will be aware that in practice, to obtain electrons moving at the same velocity, they are made to undergo a drop in potential $P$ and we have

$$\frac{1}{2} m_0 v^2 = eP.$$

Hence,

$$\lambda = \frac{h}{\sqrt{2 m_0 e P}}.$$

Numerically this gives

$$\lambda = \frac{12.24}{\sqrt{P}} 10^{-8}\text{cm} \qquad (P \text{ in volts}).$$

Since it is scarcely possible to use electrons other than such that have undergone a voltage drop of at least some tens of volts, you will see that the wavelength $\lambda$ predicted by theory is at most of the order of $10^{-8}$ cm, i.e. of the order of the Ångstrom unit. It is also the order of magnitude of $X$-ray wavelengths.

Since the wavelength of the electron waves is of the order of that of $X$-rays, it must be expected that crystals can cause diffraction of these waves completely analogous to the Laue phenomenon. Allow me to refresh your memories what is the Laue phenomenon. A natural crystal such as rock salt, for example, contains nodes composed of the atoms of the substances making up the crystal and which are regularly spaced at distances of the order of an Ångström. These nodes act as diffusion centres for the waves and if the crystal is impinged upon by a wave, the wavelength of which is also of the order of an Ångström, the waves diffracted by the various nodes are in phase agreement in certain well-defined directions and in these directions the total diffracted intensity is a pronounced maximum. The arrangement of these diffraction maxima is given by the nowadays well-known mathematical theory developed by von Laue and Bragg which defines the position of the maxima as a function of the spacing of the nodes in the crystal and of the wavelength of the incident wave. For $X$-rays this theory has been admirably confirmed by von Laue, Friedrich, and Knipping and thereafter the diffraction of $X$-rays in crystals has become a commonplace experience. The accurate measurement of $X$-ray wavelengths is based on this diffraction: is there any need to remind this in the country where Siegbahn and co-workers are continuing their fine work?

For $X$-rays the phenomenon of diffraction by crystals was a natural consequence of the idea that $X$-rays are waves analogous to light and differ from it only by having a smaller wavelength. For electrons nothing similar could be foreseen as long as the electron was regarded as a simple small corpuscle. However, if the electron is assumed to be associated with a wave and the density of an electron cloud is measured by the intensity of the associated wave, then a phenomenon analogous to the Laue phenomenon ought to be expected for electrons. The electron wave will actually be diffracted intensely in the directions which can be calculated by means of the Laue-Bragg theory from the wavelength $\lambda = h/mv$, which corresponds to the known velocity $v$ of the electrons impinging on the crystal. Since, according to our general principle, the intensity of the diffracted wave is a measure of the density of the cloud of diffracted electrons, we must expect to find a great many diffracted electrons in the directions of the maxima. If the phenomenon actually exists it should thus provide decisive experimental proof in favour of the existence of a wave associated with the electron with wavelength $h/mv$, and so the fundamental idea of wave mechanics will rest on firm experimental foundations.

Now, experiment which is the final judge of theories, has shown that the phenomenon of electron diffraction by crystals actually exists and that it obeys exactly and quantitatively the laws of wave mechanics. To Davisson and Germer, working at the Bell Laboratories in New York, falls the honour of being the first to observe the phenomenon by a method analogous to that of von Laue for $X$-rays. By duplicating the same experiments but replacing the single crystal by a crystalline powder in conformity with the method introduced for $X$-rays by Debye and Scherrer, Professor G. P. Thomson of Aberdeen, son of the famous Cambridge physicist Sir J. J. Thomson, found the same phenomena. Then Rupp in Germany, Kikuchi in Japan, Ponte in France and others reproduced them, varying the experimental conditions. Today, the existence of the phenomenon is beyond doubt and the slight difficulties of interpretation posed by the first experiments of Davisson and Germer appear to have been satisfactorily solved.

Rupp has even managed to bring about electron diffraction in a particularly striking form. You will be familiar with what are termed diffraction gratings in optics: these are glass or metal surfaces, plane or slightly curved, on which have been mechanically traced equidistant lines, the spacing between which is comparable in order of magnitude with the wavelengths of light waves. The waves diffracted by these lines interfere, and the interferences give rise to maxima of diffracted light in certain directions depending on the interline spacing, on the direction of the light impinging on the grating, and on the wavelength of this light. For a long time it proved impossible to achieve similar phenomena with this type of man-made diffraction grating using $X$-rays instead of light. The reason was that the wave-length of $X$-rays is much smaller than that of light and no instrument can draw lines on a surface, the spacing between which is of the order of magnitude of $X$-ray wavelengths. A number of ingenious physicists (Compton, J. Thibaud) found how to overcome the difficulty. Let us take an ordinary optical diffraction grating and observe it almost tangentially to its surface. The lines of the grating will appear to us much closer together than they actually are. For $X$-rays impinging at this almost skimming incidence on the grating the effect will be as if the lines were very closely set and diffraction phenomena analogous to those of light will occur. This is what the above-mentioned physicists confirmed. But then, since the electron wavelengths are of the order of $X$-ray wavelengths, it must also be possible to obtain diffraction phenomena by directing a beam of electrons on to an optical diffraction grating at a very low angle. Rupp succeeded in doing so and was thus able to measure the wavelength of electron waves by comparing them directly with the spacing of the mechanically traced lines on the grating.

Thus to describe the properties of matter as well as those of light, waves and corpuscles have to be referred to at one and the same time. The electron can no longer be conceived as a single, small granule of electricity; it must be associated with a wave and this wave is no myth; its wavelength can be measured and its interferences predicted. It has thus been possible to predict a whole group of phenomena without their actually having been discovered. And it is on this concept of the duality of waves and corpuscles in Nature, expressed in a more or less abstract form, that the whole recent development of theoretical physics has been founded and that all future development of this science will apparently have to be founded.

www.ingramcontent.com/pod-product-compliance
Lightning Source LLC
LaVergne TN
LVHW020046110826
845155LV00029B/657

* 9 7 8 1 9 2 7 7 6 3 9 8 8 *